T0234565

SpringerBriefs in Latin American Studies

Series editors

Jorge Rabassa, Ushuaia, Argentina
Eustogio Wanderley Correia Dantas, Fortaleza, Brazil
Andrew Sluyter, Baton Rouge, USA

More information about this series at http://www.springer.com/series/14332

José Borzacchiello da Silva

French-Brazilian Geography

The Influence of French Geography in Brazil

José Borzacchiello da Silva
Universidade Federal do Ceará
Fortaleza, Ceará
Brazil

ISSN 2366-763X   ISSN 2366-7648   (electronic)
SpringerBriefs in Latin American Studies
ISBN 978-3-319-31022-0   ISBN 978-3-319-31023-7   (eBook)
DOI 10.1007/978-3-319-31023-7

Library of Congress Control Number: 2016935586

The work was first published in 2012 by Edições UFC with the following title: "França e a Escola Brasileira de Geografia: verso e reverso", ISBN: 978-85-7282-512-2.

Printed on acid-free paper

This Springer imprint is published by Springer Nature
The registered company is Springer International Publishing AG Switzerland

Foreword

In 1998 Brazil was the honored guest country at the International Book Fair in Paris, and one of the round tables that was organized on this occasion by the Syndicat National de L'Édition, of which I had the honor of participating, was entitled "the eventful journey of love in the relationship between France and Brazil." It seems that this title could also be applied to this book by José Borzacchiello da Silva about the relationship between French and Brazilian geography, as evidenced in the words that he uses to describe the phases in this relationship: "approximations", "distancing", "exclusivities", "ruptures". There is actually much kindness and even much passion, at times disappointed, in the long history of this couple which inevitably has had its ups and downs.

José Borzacchiello da Silva discusses these phases in detail, offering a unique periodization in which the founding of USP in 1934 and the UGI congress in 1956 in Rio de Janeiro are key moments, and which returns to the distant origins and brings us up to the period in which he did a comprehensive postdoctoral research in France on this subject. He had by then (in 1992–1993) analyzed the 104 volumes of the *Bulletin Intergéo* published since 1966 by the CNRS (National Council of Scientific Research) and the production of French theses about Brazil and by Brazilians in France. He provides, therefore, a considerable amount of firsthand data which gives his work a notable density and which significantly supports the relevant and subtle analyses that he makes as he interprets them.

His objective is clear:

> The study of the formation of a Brazilian geography emerging from the relations established between France and Brazil allowed the learning of the level of involvement of geographers from both countries, identifying different moments according to their conjectures

The way of doing this is clearly explained:

- Enumerate the French geographers who worked on Brazil or influenced Brazilian geography by supervising theses, editing books, texts, or other activities;
- Verify, in analytical terms, the moments in which this influence peaked;

- Relate aspects of the life history of these professionals with the type of influence exercised on Brazilian geography;
- Map, in the Brazilian space, the areas which were most studied and which were the centers which established a greater geographical exchange with French universities;
- Locate in France the cities and universities which most influenced Brazilian geography;
- Locate in France the cities and universities which received the largest number of Brazilian geography researchers;
- List the main research trajectories which maintained a close connection with France;
- Identify through the titles of research by French geographers in Brazil, the level of innovation regarding theoretical, conceptual, and methodological focus;
- Verify the extent to which these innovations are inserted and assimilated in terms of Brazil.

The book deals with most of these issues but we emphasize here that one of its main conclusions is that the relationship between French and Brazilian geographers became less asymmetric. In the beginning, and for decades:

> The relationship established between the professionals of both countries was not marked by symmetry, rather the French academic tradition, associated with a long period of applying theories and methods in its expansionist phase, was guaranteed a universal character, until then unknown by the Geography carried out in Brazil

As a result,

> In Brazil, the French did their research and in the past contributed towards a geographical reading of the country, envisioning the national project from their perspective. In France, Brazilian professionals acted predominantly as apprentices of the theoretical and methodological guiding structures of this geographic school with its various lines of thought

As time went by, however, Brazil changed considerably as did the relationship between Brazilian and French geographers, which has cooled a bit,

> Geography, in line with the experience Brazil was going through of economic growth and large infrastructure projects, played a prominent role at that time. It searched in quantitative methods for possible formulas to offer spatial explanations and to deliver the expected results. At that time there was a partial rupture with the more classical form of French Geography, based on ideas of arrangement and spatial organization. This dominant practice on the part of official Geography met the interests of the military government and of the bureaucratic elite that had installed itself in the country. On the other hand, the leftwing, supported by the Church and other non-official institutions looked for theoretical references that could support their utopian realities and party proposals

José Borzacchiello da Silva notes, however, a case which crosses these diverse phases without losing his influence, that of Pierre George:

> Among the many that were here Pierre George deserves emphases as he exerted a great influence on Brazilian geography. He became an almost obligatory bibliographical reference point in geography courses, In the discussions which focus on the relationship

between French and Brazilian geography, Pierre George can be classified as an example of a situation of permanence [...] Independent of theoretical and methodological options the status of the illustrious professor remained unaltered, editing his books in Portuguese through DIFEL, European Book Diffusion, of Sao Paulo, under the auspices of the Presses Universitaires de France, in Paris and the Editora Fundo de Cultura in Rio de Janeiro. He maintained a captive audience. This stability achieved by Pierre George does not indicate in any way that the academic relationship with French scholars continued their regular path

Here we note how José Borzacchiello da Silva is capable of linking in a convincing way economic development, political situations, and their impact both on Brazilian geography and its relationship with French geography. The same applies to the following period, in which the progressive reduction of the political pressure linked to the end of the military regime permitted Brazilian geographers to considerably change the direction of their research, enthusiastically turning to the Marxist analyses which they had previously been forbidden:

The post-1978 Geography, through a significant part of the geographers' category [...] found elements for the reorganization of the science in the scarce literature based on Marxism. The movement named Fortaleza 1978 divided opinions and positions regarding ways of conceiving, teaching and applying Geography in the country [...] the period after 1978 signified a rearrangement in the map of geographic production in the country, inserted new characters in the scene, sacralised some and demonized others

We only lament that José Borzacchiello da Silva has not prolonged his analyses until the current days, we would like to know how he perceives the evolution until today, 34 years after the "turning of tables" in Fortaleza and 20 years after his research. Maybe this could be the object of a new research from both sides of the Atlantic.

Yet even as it is, this book is already fascinating (even for those who are not, as I am, one of the actors in this French–Brazilian relationship), and one of its most interesting and original aspects is that José Borzacchiello da Silva, during his time in France for postdoctoral studies, interviewed various French geographers who carried out an important role in the collaboration between the two schools of geography, always following the same questionnaire, but with a few adaptations so as to "learn the maximum about the contributions capable of answering the questions which guided the research." He has transcribed in the book the answers of his interviewees, and on occasions his own reaction—often enthusiastic—to these answers, and the additional discussion which occurred when he, always politely but assertively, made his dialogue partners complete their answers. Jacques Lévy thereby admits that "to be honest, I must say that the Brazilian Geography that I know is that which revolves around Milton Santos" and that in fact his participation in the French–Brazilian collaboration was minimal (with only two brief visits to Brazil). One of the interviews obtained an unexpected result revealing that Yves Lacoste "had no idea of his importance in Brazil, especially through his book, *A Geografia serve*" a book which was very influential in the evolution of Brazilian geography, first in the form of a pirate edition (under the military regime) and subsequently in an open form, after the return of democracy.

With Paul Claval, whose cooperation with Brazil was, and still is, important, a conclusion is imposed which is even stronger considering that his participation occurred rather late in this period, when Brazilian geography was well-established: "The relationship I had with Brazilian colleagues was always interesting for me and I have the impression of being on a terrain in which I speak with equals."

Michel Rochefort comes to the same conclusion: "there was a renovation in French geography and there is a beautiful phase in Brazilian Geography in which they are both able to come together, but now the relations are adult." Yet the most noteworthy is that this analysis comes after he has monitored a significant change which he has observed since the 1950s (at the time of the interview he had spent 27 seasons in Brazil):

> There was the infant phase when Brazilian geography was formed by French teachers, there was an adolescent phase when Brazilian geography rejected its parents and, finally, now there are adult relations when there is a conversation, there is an exchange, but neither has a supremacy over the other

Michel Rochefort carries out, however, a light critique of his Brazilian colleagues who at the time of the interview—and for many this is still true—had not dared to go beyond the limits of their own country to study neighboring countries or even more distant ones, which has meant that they could not take full advantage of their cooperation with French geographers, who have had a long history that has prepared them to analyze foreign countries without complexes and make comparisons between them: "Brazil has become closed within Brazilian Geography [when] the approaches that the French could do are comparative approaches."

On the whole José Borzacchiello da Silva makes a differentiated analysis of the French influence on Brazilian geography: "among the foreigners, there is little doubt that in Brazil, the French had and still have a noteworthy role" [...]

> The myth surrounding the quality of French Geography lasted for more than half a century. It is not important to know whether it was developed within or beyond French borders. French academics, especially those who supervised theses and coordinated laboratories were invited to Brazil to teach courses, give lectures, accompany field work and advise research groups and ministerial teams. Despite the myth and the significant reaction of French Geography when faced with Germanic Geography, it is still doubtful whether one can know if French Geography would still be able to maintain itself in a comfortable position as in previous years

José Borzacchiello da Silva leaves the door open for future discussions, with a clear sense of optimism and a proactive attitude: "the relationships between both countries should be strengthened, allowing a reciprocal exchange." Yet this optimism and this desire for a more symmetric relationship to be confirmed are based on a—justified—confidence in the maturity of Brazilian geography, which should permit it to look now beyond its own borders: "it is up to Brazilian Geography take on an important role in the explanation of the reality of Brazil, Latin America and why not, the world." French geographers will be proud and happy in accompanying them in such a praiseworthy ambition.

Hervé Théry
Director of Research at CNRS
Visiting Lecturer at USP

Preface

The relations between France and Brazil in the area of geography have always attracted my interest. My professional formation has been built with much proximity with French culture, especially geography. This book comes as a result of research developed during the time I lived at the ParisIV-SorbonneUniversity where I received the unconditional support of Prof. Paul Claval who has a profound knowledge of French geography, is a prolific researcher, and is in constant discussion with Brazilian geography. It was a privilege to enjoy this distinguished company and share theoretical and methodological discussions. I have the pleasure of being considered as part of his circle of friends.

The study of the different aspects of French geography, its schools and tendencies, the creation and diffusion of its lines of research throughout the world, especially in Brazil, led me to remain 18 months in Paris, where I had many opportunities to deepen the object of my study.

I chose courses and seminars establishing the following criteria:

- Perceive the general level of discussion in Human Geography;
- Explore the courses and seminars which worked on new themes and maintained relations with geography activities in Brazil;
- Reflect on areas which allowed the possibility of outlining and perceiving the wide spectrum of human sciences in France.

Among the general criteria I included the choice of institutions to carry out the research. I chose the Geography Institute, located at 191 Rue Saint Jacques, which hosts the geography courses of the Universities of Paris I and IV. The BoletimIntegeo facilitated my research in its archives about the activities of French geography. The library of the Geography Institute of Sorbonne was fundamental for the research to move forward.

In France I also had opportunity to collaborate with Profs. Martine Droulers, Michel Rochefort, Jacques Lévy, Yves Lacoste, Marion Aubrée, Alain Touraine, Cornelius Castoriadis, and Augustin Berque.

I emphasize the discussions made recently with deceased Bernard Lepetit. His interest in the theme increased my own commitment.

Alex Mengue, a geographer from the Republic of Cameroon, was an excellent dialogue partner. We shared rich and fruitful discussions.

Clélia Lustosa, Vanda Sales, and Maria Geralda de Almeida with their criticism and suggestions allowed the deepening of my research and the perfecting of the analysis.

I thank Eustógio Dantas, a companion in many journeys, for his insistence in publishing the book.

I thank my companions in the Federal University of Ceará (UFC, Brazil) Geography department, especially in the postgraduate program and the Metropolis Observatory, for their constant encouragement and the way that they face challenges.

I thank those I supervise, for the scientific discussion and the pleasant and jovial interaction.

I thank Prof. Hervé Théry, for the kindness in writing the preface.

Emília, my wife, a companion for all moments, who participated actively during the research period and the development of the book.

I thank National Center of Scientific and Technological Development (CNPq, Brazil) that financed the research.

José Borzacchiello da Silva

Contents

Chapter 1
Introduction

Abstract The introduction emphasizes the analysis of French geography, starting from the discussion of its theoretical presuppositions as a science, and the strong activity of French geographers in a constant search for legitimacy in relation to other fields of knowledge. It accepts the existence of a French school of geography and evidences the importance that French geography has acquired in the world, especially in Brazil. It singles out the events which strongly marked the influence of French geography in Brazil and recognizes different phases in its trajectory in Brazil, among them those of closer proximity and those of ruptures.

Keywords Geographical schools · Trajectory · Approximations · Ruptures

Pouring over satellite images, it is possible for the observer to see, perceive, and depict the Earth in its finite totality contained in the sphericity of its sides (Pinchemel).[1] A reality these days, in the recent past these images were virtual, an unattainable fiction. Today, with technological advances, everything seems possible. Is this possibility what could be called the end of geography? As a result, this classic science would become another branch of learning or maybe a technique. As a technique, it would lack the basic concern that characterizes geography and on which it is based: location, description, and comparison. Would it be the end of all possibility of discovery, observation, and analysis? Would it be the end of the secreting of emotions that makes geography a fascinating science? A science with the charm of an unveiling, a discovery, the unusual, the premeditated, and the inferred, ultimately, a special "something more" that only passionate devotees can feel… as I do.

Has the secret desire for adventure in partnership with scientific assumptions become finite?

From its beginnings until the present, geography's history has been the great motivator of its followers. The sense of daring and outrage felt in unveiling the relationship between society and nature.

---

[1]Pinchemel (1992).

© The Author(s) 2016
J.B. da Silva, *French-Brazilian Geography*, SpringerBriefs
in Latin American Studies, DOI 10.1007/978-3-319-31023-7_1

Geographical charts, terrestrial globes, facts, photographs, processed realities, and satellite images fascinate the observer. If they are a geographer, this fascination appears to increase. These are professionals who work on various levels. In their observations and analyses, they try to dominate and command the world when reaching, in terms of the production and reproduction of images, an extraordinary capacity for reduction. The much vaunted synthesis is being reached. This symbiotic wholeness of nature and society, at the level of observation and representation, is the possibility of knowledge and control. Nature blended with society shows itself to be trainable, static, indolent, tamable, and controllable.

This reality does not reduce geography to representation and observation, but without a doubt it preserves a historicity, a whole process where a restless search propelled humanity in all directions. Sailing the seas, climbing mountains, crossing deserts, and penetrating forests... In a long, slow process, the world gradually became dialectically viable in terms of macrospaces and imposed an optical rigor at the level of unknown microspaces.

Dealing with such different scales, geographical science has made a long journey.

Inserted in this process, geography had its genesis, its evolution, its anxieties, doubts, and worries. Any attempt to reconstruct the history of geographical thought could mean overlooking the epistemological discussion that underpins it and guarantees its statute as a science.

In the context of the history of this science, while searching for the origin and evolution of a branch of learning one builds up a history of knowledge. When producing this history, the narrative content and the political implications are directly linked and intertwined with the narrator. History depends on the expositor.

Concerning the relationship between science and conception of the world, Gramsci affirms that:

> Placing science at the foundation of life, making science the par excellence conception of the world, which frees the eyes from any ideological illusion and faces man with reality as it is, means to fall back on the concept that the philosophy of praxis needs philosophical props outside of itself. However, in reality, science is also a superstructure, an ideology. Nevertheless, it is possible to say that science has a privileged position in the study of superstructures as its reaction to nature has a particular character; it has a greater extension and continuity of development, notably after the eighteenth century when science became a superstructure. This is also demonstrated by the fact that it has had entire periods of eclipse when it was obscured by a different dominant ideology, religion, which affirmed that it had absorbed science. Thus science and the Arabic know-how were perceived by Christians as pure witchcraft. In addition, despite all the scientists' efforts science never presented itself as a naked objective notion. It always appears covered by an ideology. In concrete terms, science is the union of an objective fact with a hypothesis, or system of hypotheses that overcome the mere objective fact. (Gramsci)[2]

---

[2]Gramsci (1987, pp. 70–71).

Geography is no exception to the rule. While the historical process of scientific knowledge confused itself with ideology, it remained at the service of power and was cloaked in an idealistic romantic vision forged in the West, especially Europe. As a forum for science in the construction of knowledge, geography has been sufficiently instrumentalized to cope with ideological discourses elaborated by different social actors.

In humanity's accumulated experience, the Earth as a planet, apparently free from more complex and consequential theoretical questions, reaches a radical change in its emblematic, that is, from the compartmentalized and distant Earth that separates and isolates individuals and societies in their relationships with its constituent elements: earth, air and, water, and in their form and distribution: continents, seas, oceans, and atmosphere to the majestic observation of the planet as part of the totality of the universe obtained from images transmitted by satellites.

From the horizontality that separates, compartmentalizes, isolates, and camouflages the planet's visibility, creating deformations in perspective, to the conquest of clear vision obtained in the altitudes and the verticality of aerial photographs of the globe, right through to the universe of images transmitted and retransmitted by powerful and complex devices that have been developed and perfected by an advanced technology where the virtual image is confused with the real one.

Humankind, accepted and conceived as a subject of the planet, is allowed to isolate itself, to become removed from the Earth as an objective universe which is external to humanity. This breaking of distances, this static dynamic articulates subjects/actors increasingly close/distant in their digitalized homes in technical-scientific-computerized spaces (Santos).[3] The flows and networks connect all the spaces, reducing them to observable images on "magic" TV and computer screens. Interminable journeys conduct the subject to a collective universe where interaction takes place through connected networks.

Nowadays, the concept of the world distorts the current concepts of limits, borders, and barriers. National minorities insist upon forming their own states. New forms of federations concentrate lands, unite countries, and experiment with unified languages in the political, economic, and cultural sectors. At the same time, nations and federations are fragmenting, creating competition. Wars and struggles support dreams, maintaining hegemonic groups and fueling plans of conquests.

This drama of the dialectical game, this interaction on differentiated scales that integrates or alternates macro and micro in a continuous affirmation and negation of changing realities is, in the last instance, the first rationale of geography. Its scientific status involves a revelation, the discovery of the world in the component pieces of a mosaic. These are general or regional interpretations, visions centered on parts of the world that instigate policies for control and modeling. This often intentional compartmentalization of the whole justifies the adoption of colonialist policies in territorial management. Confirming this practice, central governments have drawn up their plans to expand and exploit land and people based on the views

---

[3]Santos (1991).

of authoritative persons. Vidal de la Blache, the eminent French geographer, makes the following consideration:

> We should congratulate ourselves because the task of colonization that was the glory of our epoch would be merely a disgrace if nature could have established rigid limits, instead of leaving a margin for the action of transformation or reconstruction that is within the power of man. (Vidal de la Blache)[4]

On the same note, based on the text above Demangeon affirms that: "Black Africa still offers wonderful prospects to European colonization"(Demangeon).[5]

Colonial geography became prominent for the French colonial empire. As academic knowledge and as a school discipline, geography grew considerably until the beginning of the twentieth century in France. According to Droulers (1991),[6] in 1923, Albert Demangeon and Emmanuel de Martonne founded the geography Institute of Paris, which had days of glory. The degree (*license*) and specialization (*aggregation specialisee en geographie*) courses were only created in 1941. Referring to Colonial geography as the base of French geography during the nineteenth century, Bruneau (1989)[7] states:

> Its purpose was to supply detailed knowledge of the colonized environment and territories in its possession in rational terms.

As a counterpoint to this classical image of geography built on its worldview, there is a postmodern geography of our times,[8] whose main territorial space/society ratio seems to escape analysis like a handful of sand running through ones fingers.

Geography insists on seeking its identity. The spatiality of the contemporary world and its reducible character stemming from the transformation of the real by the image places geography in an uncomfortable position among the other human sciences.

"Which mirror did I lose my face in?" In her poem "Retrato Natural" (1987)[9] Cecília Meireles explores the tragic discovery of the loss of her features, the ones that guarantee her identity. In which mirror/images has geography constructed its face, at least the one most commonly known?

In this game of images, reflections, mirrors, and identities, geography tries its luck, its path. It seeks to construct a profile that will guarantee reliability and

---

[4]de la Blache (1989, p. 15).

[5]Demangeon (1947), in: Santos, M. op. cit. p. 15.

[6]Droulers (1991).

[7]Bruneau (1989), in: Droulers M. op. cit. p. 35.

[8]Baudrillard's description of American space as a counterpoint to European space provides this idea of a post-modern reading of geographical space. The following passage suggests this situation: "The sunsets are rainbows that last an hour. The seasons there do not make sense: springtime in the morning, summer at midday and the desert nights are cold without ever being winter. It is a type of suspended eternity or the year is renewed every day." In: Baudrillard (1986, p. 117).

[9]Meireles (1987, p. 84).

permanence. Its existence rests on the possibility of becoming increasingly useful, reflective, and practical when faced with the changes of a new age.

Technological innovation happens with overwhelming speed. The large economic groups that control the creation and production of innovations wage an unprecedented battle for profit. Autophagic policies are adopted so that new models often undermine those only recently offered to the public. The insane greed of the market in its desperate search for profit surrounds humanity converting them into consumers everywhere, all the time.

In the name of modernity, this process advances more and more. There is no limit! Nearly new products are thrown on the scrapheap when they are overtaken by products that have just come off the production line. The life span of products has been reduced. Each newly launched product has a type of *announced death*; it is a prototype that, unlike Niccolo Pisano's "Annunciation" (sixteenth century), foretells its own death. It is regrettable... but this change and speed does not affect the world synchronistically. Different spaces and times create a logic where hunger and plenty, disease and wellness, and deficit and surplus coexist among other antipodes in a symbiosis based on injustice and inequality.

Geography built its history on this long and difficult path, gaining renown and respectability. There were times when intellectuals met at geographical societies such as the ones in Paris, created in 1821, Berlin in 1828, and London in 1830. An air of science gave geography the charm of method, theory, and analysis. The empirical basis, the expeditions, the journeys, and field studies shrouded this area of knowledge with mystery, the mystery of adventure, of the new, of facing the unknown. Geography gained prominence, more breathing room. It adopted instruments and became knowledge at the service of power (Santos 1978; Dresch 1948). Addressing this issue in the context of the discussion of geography's *utility*, Lacoste was emphatic:

> Taking as a starting point that geography serves first and foremost to wage war is not to say that it only serves to conduct military operations. It also serves to organize territories, not only to predict which battles can be waged against such and such an adversary, but to better control the people over whom the State exercises authority. Geography is above all strategic knowledge linked strictly to a set of political and military practices... (Lacoste)[10]

Geography has carried this taint for a long time and may still be carrying it. Quaini (1978)[11] denounces its lack of usefulness as teaching material, as does Lacoste (1982)[12] when he identifies a "teacher's geography" and another that he calls "geography of larger states."

In the search for legitimacy and acceptance, geography has had different approaches. Some were confused with techniques, such as the new geography that

---

[10]Lacoste (1982, p. 7).

[11]Quaini (1978).

[12]Lacoste Y. op. cit.

installed the so-called quantitative revolution in Brazil. Referring to this deployment Monteiro[13] affirms:

> After advice given by M. Rochefort to the CNG during transformations in 1966, one of the sensitive changes in the orientation of the new IBGE Foundation and its new IBG was the reopening of Anglo-Saxon geography. On the threshold of the transition to the second term (1967-1968) there was the effective introduction of quantitative techniques and "theoretical" concerns through visits from Gauthier, Cole and Berry.

In other approaches, geography was more of a method, such as the regional method proposed by Hartshorne.[14] According to Claval, this was a type of reaction to Fred Schaefer who contested that geography was a unique science in an article entitled "Exceptionalism in geography." Claval[15] is categorical:

> Thus, Schaefer's article is indirectly the origin of the publication *Perspective on the nature of geography*, which in addition to being a response by Schaefer, is a counter-manifesto.

This approach as a school gave geography features geared toward certain national models. This led to the characterization of a German School, a French School, a Swedish one, etc.

The immediate acceptance of the existence of a national school can often mean a reductionism. A closer analysis of what is conventionally called a school can reveal different groups which may be well or less defined and which in many cases oppose each other. However, what gives visibility and characterizes or configures a school is a group united around certain principles or doctrines. For the purposes of analysis, this research understands a school to be the national group and attempts to verify whether Brazilian geography has constituted or constitutes itself as a school.

According to Capel (1999):

> The existence of a specialized scientific community models the thought of its members and, with time, gives birth to what has been called *styles of thought* which determine the choice of scientific problems, the questions which are asked, guide the observations, establish the working rules and even predetermine the vocabulary to be used. What in many scientific disciplines has been called a point of view may be little more than the application of the style of thinking pertaining to the community (Capel 1999).[16]

At the same time, the development of the work and reality reveal the issue and show evidence of the blending of principles, concepts, and methods in global terms, which hampers the construction and definition of local identities. The globalization of the economy, megacities, the revolution in transport and communications, and the most varied forms of languages mix and overlap ideas, making it difficult for the analyst to prepare certificates of originality. Publishers distribute their catalogues worldwide, universities intersect through exchange programs. Computers connect and interact in specific systems organized in the form of networks. Society,

---

[13]Monteiro (1980, p. 27).

[14]Hartshorne (1939).

[15]Claval (1976, p. 185).

[16]Capel (1999, p. 20 and 21).

landscapes, the environment, space, and territory appear on the big screen. Nevertheless, it is known that in the social relations of production that structure and streamline the market, the system of exchanges and interchange generate a situation of inequality and exclusion.

Brazil and other peripheral countries are part of the excluded and have serious difficulties in projecting their conception of science, and more specifically, geography. As in other counties, at first there is not a single school. Even inside universities, there are groups of affinities that are opposed as to their concepts, doctrines, principles, and the underlying philosophical basis that sustain their positions in the context of geographical science.

In this work, the Brazilian school will be treated as the systematized set of knowledge, known and identified as geography.

This knowledge produced in the country is sometimes limited to its frontiers, as if Brazilian geography were only capable of generating interpretative knowledge about its own reality.

The Institute of geography, in Paris, was this geographers' privileged location. The rest of the research was carried out at the Ecole des Hautes Etudes en Sciences Sociales (EHESS) an institution that has great international prestige as it brings together professionals from all the Humanities and at IHEAL, the Institut des Hautes Etudes de l'AmeriqueLatine.

Alongside developing the research and attending courses, I joined two groups that brought together French and Brazilian researchers who organized seminars and discussions on the Brazilian reality.

- GRUPO BRASIL, facilitated by the geographers Martine Droulers and Bernard Bret from the CREDAL (Centre de Recherche et de Documentation sur l'AmeriqueLatine), URA from the CNRS, associated with the Institut des Hautes Etudes de l'AmeriqueLatine.
- GROUPE DE REFLEXION SUR LE BRESIL CONTEMPORAIN facilitated by Ignacy Sachs and Marion Aubree from the Centre de Recherches sur le BresilContemporain da Maison de Sciences de l'Homme.

The triangle composed by the Institut de Geographie, the Institut des Hautes Etudes de l'AmeriqueLatine and the Ecole des Hautes Etudes en Sciences Sociales da Maison de Sciences de I'Homme were a rare opportunity to *look, compare* and *question* the current state of French geography and its repercussions in Brazil. Inserted in the historical context of the relationship between the two countries, the research started to acquire a clearer shape being able to trace the trajectory of Brazilian geography, especially through a return to its origins. It was thus possible to recover the first academic contacts established between the two countries from a survey of the scientific production of Brazilians in France and the French professionals who elected Brazil as a theme and a field of research.

The study of the formation of a national geography stemming from the relations established between France and Brazil allowed the discovery of the level of involvement of geographers from both countries identifying different moments according to their conjectures. The intention of building a profile of Brazilian geography in keeping with the influences of the French geographical school allowed us to: enumerate the French geographers who have worked on Brazil or influenced Brazilian geography, whether through the supervision of theses, editing books, texts, or other activities, analytically verify the peak moments of these influences, relating aspects of these professionals' life histories with the kind of influence they exerted on Brazilian geography; map the areas in Brazilian space that have been studied; map those centers that have established the most geographical exchanges with French universities; locate in France the cities/universities that have received the highest number of Brazilian researchers in the field of geography; list the main lines of Brazilian research with close links to France; identify the level of innovation in the theoretical–conceptual and methodological focus through the titles of studies carried out by French scholars in Brazil and verify the extent to which the insertion and assimilation of these innovations occurs in terms of Brazil.

Given its historical–analytical character, the definition of the methodological procedures and scope of the research demanded a survey of primary and secondary sources, interviews, and the production of charts and tables.

The main source of primary research was the series of 104 volumes of the BOLETIN INTERGEO, published since 1966 by the CNRS, which records the main aspects of the development and practice of French geography. The structure of this research satisfactorily addresses the questions that guide it. Intergeo deals with aspects of the formation of tables and publishes lists of professors and researchers at French universities, providing information regarding the most important events of the geographical community in that country. It also provides details of geography theses in France at any given period, regarding publication and/or defense as well as future defenses. Being able to rely on this information made it possible to obtain knowledge of the evolution of French geography, identifying its main producers and exponents during this period. Another interesting aspect was the possibility of verifying the evolution of themes and correlating the subjects researched by Brazilians with data on the country's reality. A thorough consultation of the 104 volumes of this report and the registration of the pertinent data was supplemented by surveys of secondary sources covering the period 1934–1966. Interviews with professors, technicians, and administrators provided the remaining information to fill the frame of reference out.

The survey provided a list of French professors and researchers who have studied or worked in Brazil, which enabled the French centers generating research on the country to be identified immediately.

The main topics related to the changes undergone in Brazilian spatiality revealed French geography's constant attempt at becoming up to date, as an important field of knowledge.

Contemporary themes and the emergence and application of new concepts revealed a period of great effervescence in geography when the terms space, organization, *aménagement*, territory, environment, and *médience*, etc., were highlighted in the geographical production of the country and were soon incorporated by the geography being constructed abroad. The community organized around a school would have a strong directive power in terms of thinking about science based on its organization and its presuppositions. For Bailly and Ferras (1997),[17] epistemology in its etymological sense is approached as a theory of science, as a dynamic of thought and a scientific discourse. Epistemology seeks three objectives: an objective of knowledge of the dominant thought, that is, the research of a problematic issue or of the greatest problems; a methodological objective to understand the modalities of acquisition and organization of the knowledge which will be used; an objective of clarifying the steps to be taken to organize thought, from the collection of data to the procedures put in place to control the results.

Epistemology acquired its scientific status following the trajectory of the philosophy of science, stemming from Descartes' works, *Discourse on Method*, of 1637, and *Essay on the Philosophy of Sciences* of Ampere, of 1860.

In Brazil, the theme of the epistemology of geography attracted stronger interest from 1978. Milton Santos was the pioneer with his book *Por uma geografia nova* (1978) in which he affirms:

> [...] since the foundation of what has traditionally been called scientific geography, at the end of the nineteenth century, it was never possible to construct a set of propositions based on a common system and linked by an internal logic. If geography was not capable of overcoming this deficiency it is because it has always been more concerned with a narcissistic discussion about geography as a discipline rather than concerning itself with geography as an object. Always, and still today, there is more discussion about geography rather than about space, which is the object of scientific geography.

As for the production of Brazilians in France, the survey showed the most studied areas, the most requested Universities and the most sought after advisors. Unfortunately, the data provided by BOLETIN INTERGEO do not give precise information regarding the year of the defense of theses of those enrolled in postgraduate courses or these students' area of knowledge.

This last aspect prevented comparisons being drawn between the number of geographers and other Brazilian professionals who did their theses in France.

Study trips to Brazil by French scholars were another topic focused on in the research. There was one period in which they were frequently in the country.

As for visits to France by Brazilian geography researchers or professors as guests of that country, the survey records a rather unequal treatment as can be expected in exchanges of this nature. IHEAL—the *Institute* of Latin American Studies is the study and research center that invited the most Brazilian scholars at a time when France had a strong interest in studying and understanding Brazil.

An examination of the courses offered by French universities indicates the types of themes related to determined professors. This survey confirmed a strong interest in themes linked to the development such as demographic growth and Third World.

The material obtained through interviews was a valuable source to understand the presence of French thinking in Brazilian geography. Of all the French professors and researchers who in some way influenced Brazilian geography, we selected

---

[17]Bailly and Ferras (1997, p. 6).

some professors who gave courses or carried out research in Brazil; supervised theses of Brazilians in France; authors of Geographical books or texts that have influenced Brazilian geographical thought, whether or not they know the country.

The interview script was the same as the questionnaire sent to the selected professors. In some cases, depending on the interviewee, there were adjustments to maximize the contributions capable of addressing the issues that guided the research.

On the whole, the material obtained during the research allowed a broad view of the presence of French thought in Brazilian geography and provided clues to the preparation of this research.

Among Brazilian teachers, I interviewed Milton Santos and Pedro Pinchas Geiger, who contributed toward piecing together information on the France—Brazil relationship in the period prior to 1966, the date the INTERGEO newsletter began.

Nowadays France has a much larger interest in Brazil. There is a renewal in the group of professors and researchers interested in studying and understanding our country. Clearly, there has been a change in attitude. The Brazil of the 1990s bears few traces of the country which was known by the exponents of French geography. The XVIII Congress of the International Geographical Union, held in Rio de Janeiro in 1956, was the event which marked Brazilian geography permanently. It was the greatest demonstration of the capacity to read and interpret facts of a geographical nature, from the Brazilian reality. Rigorous in theory and method, Brazilian geography revealed itself to be vigorous with formidable official support, in this case the IBGE. The foreigners who visited the headquarters of the National Council of geography in this year were impressed with the dynamism and the diversity of Brazilian geographical production. The event was significant:

> For the first time the I.G.U. held an international conference in the tropics and the southern hemisphere, which attracted a French delegation consisting of first-rate geographers led by MaximilienSorre.[18]

Beyond the activities developed eighteen excursions were prepared, traveling over the vast extension of the Brazilian territory, one occurring before and one after the XVIII International geography Congress. The guides for the excursions explored the most important geographical aspects of the regions visited. All were published by the National Council of Geography. Top-level professionals were gathered to publish these guides and assure the success of the event. The Portuguese version began to be edited in 1957. The excursions with their corresponding guides were the following: (1) central-western plateau and the Mato-Grosso swamp region of Fernando F.M. de Almeida and Miguel Alves de Lima (1959); (2) the metallurgical zone of Minas Gerais and the Rio Doce Valley of Ney Strauch (1958); (3) the coffee march and the pioneer frontiers of Ary França (1960); (4) the Paraiba Valley and the Mantiquera range and the area around São Paulo, Aziz NacibAb'SaberandNiloBernardes (1958); (5) coastal plain and sugar growing

---

[18]Valverde (1984, p. 83).

region in the state of Rio de Janeiro of Lysia Maria Cavalcanti Bernardes (1957); (6) Bahia of AlfredoJosé Porto Domingues and Elza Coelho de Souza Keller (1958); (7) scenes from the northeast in Pernambuco and Paraíba of MarioLacerda de Melo (1958); (8) Amazon of Lúcio de Castro Soares (1963); and (9) the southern plateau of Brazil of Orlando Valverde (1957). These were published during the event, in English and French, being one of the best indicators of the Brazilian capacity to produce geography in the second half of the twentieth century, when there were still few graduate courses in geography in the country. The French delegation was the largest.

If today there is improved French interest in Brazil, the reciprocity is not perfect. France remains one of the most popular countries for Brazilians to do their post-graduate courses. The number of Brazilians enrolled in courses in the European country is reasonable. The translation of texts and publications by French authors in Brazil is also relevant.

The circumstances of geography in both countries have taken different paths and experienced distinct realities, which has imposed new elements for reflection. In France, after the great post-Marxist theoretical crisis, the situation in Eastern Europe and the construction of a federative Europe has turned out to be a European geography seeking to understand themes that have not yet been studied. Geographies with a national focus have acquired new contours with Europe's new design. The European Community and Europe in a broader sense have strongly propelled French geography to understand the continent's new configuration. This attitude does not invalidate their traditional practice—France and the World; the World being understood as Black Africa, the Maghreb, Latin America, and Southeast Asia.

Brazil, in turn, in the search for a National geography, faces a deep, long crisis that has imposed a search for new frameworks capable of explaining this reality. Seen in a historical context, this difference in posture and interests highlights the geographies of both countries.

French geography has a great prestige and international recognition but lived its golden *years* from the beginning of the twentieth century with its heyday in the 1950s and 1960s. Brazil while being influenced by other geographical schools such as the American, German, and English schools became established as a country under the strong influence of French geography.

The creation of the degree course in geography in Brazil at USP—the University of São Paulo—in 1934, the same year the University was created, led to the formation of national set of teachers that gradually helped and replaced foreign teachers as their activities in the country came to an end.

Gradually, Brazil sought its autonomy through the creation of new geography courses (Rio de Janeiro, in 1935) and the creation of the IBGE (the Brazilian Institute of geography and Statistics) in 1937. The IBGE played a fundamental role in defining, shaping, and understanding Brazil's spatial reality. The organization qualified various generations of geographers to an excellent technical standard, becoming one of the main "schools" of geography in the country.

The research enabled a clearer understanding of the trajectory of Brazilian geography. The return to its origins, the reading of accounts of journeys, access to a vast bibliography, and carrying out several interviews and contacts with professors and researchers who had a big influence on Brazilian geographical thinking were, in short, revealing.

The importance attached to French scientific production and the process of maturity achieved by our intellectuals goes without saying. Sergio Miceli, in his book *Les intelectuelset le pouvoir au Brésil*, comments on the issue:

> The emergence of a particular category of intellectual, the professional researcher, was consolidated in Latin America after the Second World War, especially in the 1960s. Accordingly, the birth of the researcher and the university student is very recent; one generation at the most. The role played by this first generation of researchers, born between 1945 and 1960, considerably modified the relationships established in the years 1920-1945 between intellectuals and scientific research on one side and intellectuals and power on the other. The analysis of issues posed by these new intellectuals regarding the big transformations that their societies had undergone in those last 30 years, seemed like a touchstone for projects establishing new research in a cooperation between France and Latin America.[19]

This statement is true in the Brazilian case. The quality of our professionals and the intellectual influence they started to exercise in taking important public policy decisions in Brazil cannot be denied. In the specific case of geographers, the recognition that they gained is representative. Beyond the quality that they displayed in their work posts various Brazilian geographers acquired an international reputation. However, even when this quality is known and recognized, French supremacy continued evident. Aspects linked to quality and the ties established between France and Brazil were cemented in such a way that there was an almost constant presence of French scholars in Brazil and Brazilians in France.

The relationship established between professionals in both countries was not marked by symmetry, on the contrary the French academic tradition associated to a long period of applying theories and methods in their expansionist action, guaranteed them a universal character, until then unknown by the geography which took place in Brazil. In Brazil, the French did their research and in the past contributed toward a geographical reading of the country, envisioning the national project from their perspective. In France, Brazilian professionals acted predominantly as apprentices of the theoretical and methodological guiding structures of the geographical school with its various lines of thought. There are few cases of Brazilians who taught in France and when they did it was mainly to talk about Brazil, the tropical world, hunger. There are few cases where they teach or taught general or European themes to the French.

In research carried out in 1992 through CAPES and the CNPq with 635 Brazilian students in France that was answered by around 50 % of them, questionnaires regarding personal evaluation showed the following results, among others: student recipients of grants ranked the guidance they received as weak (28 %), good (43.6 %), and very good (45.2 %). As for the advisor's interest in the student's research topic, 20 % considered it regular, 32.5 % strong, and 45.2 % very

---

[19]Miceli (1985, pp. 28–30).

strong. In the same survey, 92 % assessed the faculty level as good or excellent and 94.7 % deemed the level of research conducted in their departments to be good or excellent. 91.7 % would recommend their supervisors, and 92.7 % would recommend their study departments to other Brazilian scholarship students.[20]

There is no specific data for scholarship students in the area of geography. However, the results reveal a certain uniformity if the diversity of the areas represented by the answers is taken into consideration.

Earlier my research had already raised new interests requiring development, including those relevant to the return of students and professionals who opted to do their postgraduate studies in France. The CAPES/CNPq survey data sharpened my scientific curiosity and provided an incentive to continue, now with another approach and other characteristics, as well as new objectives and aspects of the course of France–Brazil relations.

This new path involved learning about the professional situation of geography in Brazil as clearly as possible, for the purpose of analysis. In other words, on returning to Brazil, what is the contribution of the professional who has previously studied in France? An analysis of the times before and after their stay in France emerges as a kind of assessment of the performance of the geographers working as technicians or teachers after returning to Brazil. The guiding questions were as follows:

• Is there a substantial change in these professionals' practice after their time in France?
• Did/Do they have an innovative function in Brazilian geographical production?
• Have they taken on significant positions at various levels of public or private administrative management in the country?
• Have they had a strong influence on public policy?

What could be observed is that the intellectual leadership of professionals in the area of geography comes upon those who completed doctorates overseas. In the case of France, this is evident. This whole picture engendered the need to rethink Brazilian geography, seeking to understand its essence and new possibilities. This book recovers knowledge about the origins and pathways of Brazilian geography, insofar as they relate to the influence exercised by the French. The thesis itself does not exhaust the subject, but allows a deeper understanding of the relationship between the two countries from the geographical perspective. It certainly provides an immediate response to the initial enquiries and assumptions directing the research.

It is apparent that the situation in recent years has outlined new power blocs, changed the map of the world, and established new relationships. In this context, old and new partnerships maintained by Brazil with so-called core countries require more accurate interpretations to monitor the performance of professional sectors in

---

[20]Published by Nouvelles APEB, Informativo da Associação dos Pesquisadores e Estudantes Brasileiros na França no 7 dez 1992, Paris.

the country and evaluate the performance of a sector of scientific knowledge of the country.

French geography established itself from the country's colonial activity. In a different historical context, it expanded under the label of Tropical geography with the unfolding of colonial geography and later on adjusted itself to the framework resulting from the Second World War, becoming a Third World geography. Today, the reality is a world integrated by the market. Given this new reality, the question arises regarding the extent to which French geography has generated knowledge able to explain it?

During the second half of the twentieth century, Brazil joined the group of countries with (late) development possibilities. In the 1970s, after the military coup, the country's modernization process involved a search for neo-positivist mathematical models capable of explaining the social reality in the country. Geography, in line with the experience that Brazil was going through of economic growth and large public infrastructure works, played a prominent role at that time examining quantitative methods for possible formulas to deliver spatial explanations and the expected results. At that time, there was a partial rupture with French geography in its most classical form based on the ideas of arrangement and spatial organization. This dominant practice on the part of official geography met the interests of the right wing and of the bureaucratic elite that had installed itself in the country. On the other hand, the left wing, supported by the Church and other non-official institutions, looked for theoretical references that could support their utopian realities and party proposals.

The post-1978 geography, through a significant group of active geographers leading the AGB (Association of Brazilian Geographers), found in the scarce literature based on Marxism elements for the reorganization of the science. The movement which was called Fortaleza 1978 divided opinions and positions regarding the way of conceiving, teaching, and applying geography in the country. Rebelling against the hierarchical structure of the AGB various students and geographers proposed other directions to the General Assembly at the end of the Third National Meeting of Geographers. The conflicts were not easily solved and maybe were never solved. In concrete terms, there was a renewal movement in geography marked by the publishing of books and the writing of articles of a Marxist perspective, with a strong social content, demanding a position from geographical professionals when faced with reality. The post-1978 period meant a rearrangement in the map of geographical production in the country, inserting new characters and leading others to be treated as sacred and others as demons.

In this context, with his book *La geographieçasertd'abord a Caire la guerre*, Yves Lacoste[21] became a kind of guru for a whole generation of Brazilian geographers. In the situation resulting from this radical change in Brazilian geography, what is the balance of the development of geography in this period through an analysis of relations between the two countries? Did Brazilian geographers at that

---

[21]Lacoste, Y. op. cit.

time produce theses with a Marxist content in postgraduate programs in France? Were there any Marxist geography books in publication phase at that time in that country?

Colonial practice, a characteristic trademark of French geography, made it the last branch of the humanities to suffer the effects of Marxist theoretical foundations. The crisis in Marxism in the 1980s imposed a search for new paradigms. Had French geography found new paths or was the publication of Lacoste's book inserted in a wider context of possible explanations of French reality?

The affirmation of French geography as a science with pedagogical content resulted in the creation of geography courses in various University centers. This expansion promoted the emergence of specialized centers, for example: Alpine geography at Grenoble, Maritime and Fishing geography at Brest, the geography of the Maghreb at Montpellier, and geography of the Tropical World at Bordeaux. This dispersal and specialization of the teaching centers led to great changes in French geography.

Regarding geographers, their areas of specialization form different groups, often coinciding with the subsections of a country or of a group of countries. Latin America becomes divided into Argentina under the influence of Pierre Denis, Brazil with the fundamental figures of Pierre Monbeig and Pierre Deffontaines, Mexico with Bataillon, Peru with Claude Colin Delavaud, and among others.

These master's disciples continued and continue working on approaches based on their formulations, making necessary adjustments. Many of them elected, like their masters, those countries as preferred areas of research; although in the case of Brazil, there is a visible decrease in the area covered by research, which is understandable considering the changes occurring in the country and the increase in the complexity demanded of each researcher. This means that HérveThéry researched Rondonia; Yves Leloup, Minas Gerais; Raymond Pebayle, Rio Grande do Sul, Martine Droulers, Maranhão, etc. Unlike their masters, who had more general approaches to the country, these researchers carried out studies in rural/urban settings.

Another piece of evidence is the reduction in the number of researchers linked to Brazil and other Latin American countries. In relation to the analysis of Brazilian reality, this reduction could be indicating that a certain autonomy has been reached. Does this also indicate that Latin America has ceased to be a center of interest for French geography? Today, French geographers, present on missions to Brazil, are mostly linked to research organizations independent of universities.

In recent decades, French geography appears to have changed its focus turning to a broader debate capable of inserting it in academic circles with greater consistency. Some more experienced geographers enjoy prestige while among the younger ones many work on new formulations. The historical RECLUS Group from Montpellier, where the M.G.M. (Maison de la Geographie de Montpellier) was a source of innovation in French geography and a reference for the world, has changed the course of its analysis completely adjusting itself to the new informational languages, converting itself into a place where innovation in cartographic representation and geographical analysis of greater complexity is located and is pointed to as a source of innovation in French geography.

Due to the close links of Brazilian geography with French geography, various members of the new dominant currents in Brazil have put the traditional affiliation into doubt.

Having exercised a strong influence on the Brazilian technical and scientific training, French geography contributed to various programs and activities in the country, causing a widespread dissemination and assimilation of new concepts such as *amenagement du territoire, ville-moyenne, metropole d'equilibre*, and technopole. Today, another reality prevails. Geography has become consolidated in Brazil, being established in various centers with the creation of graduate and postgraduate courses.

Concerning the advance of Brazilian geography, Milton Santos in the early 1990s already affirmed that the country presented a counterpart:

> It is time for us to explain what we consider Brazilian geography to be. This is not to propose that the country's borders should be closed with a kind of sanitary isolation cord to prevent people and ideas developed abroad from settling and having influence here. Not at all. There is another objective. It is a case of helping the country become a mature nation from an intellectual and cultural point of view. A country that is incapable of generating its own ideas is fated to become a dependent country, or even not be a country. But building a Brazilian geography also means building geographical thinking, which although born in Brazil, is universal.[22]

This theme has provoked a wide debate. If we remain dependent and fail to reach a point characterized by the construction of our own discourse, all we are doing is importing formulas and models and camouflaging solutions applied elsewhere.

The construction of knowledge presupposes a slow process. Geographical knowledge is conceived and produced from original ideas that characterize our capacity to impose challenges on ourselves. It is important to remember the importance that Milton Santos attaches to the exchange of ideas and living with foreign professionals.

Milton Santos insisted in the construction of a Brazilian geographical thinking. His pioneer approach appears in the book, *Por uma geografia nova,* of 1978. He undertakes a severe criticism of the path that geography was taken affirming:

> [...] since the foundation of what has traditionally been called scientific geography, at the end of the nineteenth century, it was never possible to construct a set of propositions based on a common system and linked by an internal logic. If geography was not capable of overcoming this deficiency it is because it has always been more concerned with a narcissistic discussion about geography as a discipline rather than concerning itself with geography as an object. Always, and still today, there is more discussion about geography rather than about space, which is the object of scientific geography.[23]

---

[22]Santos (1982, p. 217).

[23]Santos (1978, p. 2).

The construction of knowledge in the context of scientific production is the expression of the formation of specialized groups formed in environments suitable for reflection, analyses, and criticism. Referring to scientific knowledge, Pinto (1969) states:

> As a process and essentially historical and progressive the scientific knowledge of any given moment is the premise of the scientific knowledge of the subsequent moment. Being methodical, it is acquired voluntarily through *rules* to explore the objective, physical and social reality that condition the nature of the results obtained.[24]

The French environment announced innovations in the theoretical–methodological field. According to Pinchemel (1992)[25], the new geography resulted, necessarily, in a change in concepts. In the concepts of Classical geography, regions, especially environment and scenery, did not resist the changes of perspective and methods and were substituted by a single concept, space.[26]

References

Bailly A, Ferras R (1997) Éléments d'épistémologie de la géographie. Armand Collin, Paris

Baudrillard J (1986) Amérique. Grasset et Fasquelle, Paris, 123 p

Bruneau M (1989) Les enjeux de la tropicalitt. Masson, Paris

Capel H (1999) O nascimento da ciênciamoderna e aAmérica. Editora da Universidade Estadual de Maringá, Maringá

Claval P (1976) Essai sur L'Évolution de la géographie humaine. In: Annales Littéraires de L'Université de Besançon. Les Belles Lettres, Paris, 201 p

de la Blache PV (1989) Géographie Générale, *Annales de Géographie*, no 38. In: Santos M (ed) Por UmaGeografia Nova. S. Paulo, Hucitec, 1978

Demangeon A (1947) Traite de Geographie. Amland Colin, Paris

Dresch J (1948) La prolétarisation des masses indigénes en Afrique du Nord. Chemins du Monde, 57 p

Droulers M (1991) L'école française de geographie. In: Monbeig P (ed) Un geographe Pionnier. IHEAL, Paris

Gramsci A (1987) Concepção Dialética da Historia. Civilização Brasileira, Rio de Janeiro, 7a. ed

Hartshorne R (1939) The nature of geography. In: Annals Association of American Geographers XXIX

Lacoste Y (1982) La Géografphie, ça sert d'abord, à faire la guerre. Maspero, Paris

Meireles C (1987) Viagem. In: Cecília Meireles Obra Poética, Rio de Janeiro, Editora Nova AguilarS/A

Miceli S (1985) Les intellectuels et Ie pouvoir au Brasil (1920/1945), Paris, Ed. de la Maison de Sciences de l'Homme, 1981. In: Chocchol J, Martiniere G (eds) L'Ammque Latine et Ie latino-ammcanisme en France. L'Harmattan, Paris

Monteiro CAF (1980) A Geogra a no Brasil (1934–1977): Avaliação e Tendências. Instituto de Geografia, Universidade de São Paulo-USP, São Paulo, 155 p

---

[24]Pinto (1969, p. 31).

[25]Pinchemel (1992, 1132 p).

[26]Pinchemel (1992, 1132 p).

Pinchemel P (1992) L'Aventure géographique de la terre. In: Encyclopédie de Géographie. Economica, Paris

Pinto AV (1969) Ciência e Existência. Paz e Terra, Rio de Janeiro, 537 p

Quaini M (1978) Marxismo e Geografia. Paz e Terra, Rio de Janeiro

Santos M (1978) Por umageografia nova. Hucitec, São Paulo

Santos M (1982) O Pensamento Geográfico ea Realidade Brasileira. In: Santos M (ed) Novas Rumos da Geografia Brasileira. Hucitec, São Paulo

Santos M (1991) Flexibilidade Tropical. In: Arquitetura e Urbanismo no 38, out/novo 1991

Valverde O (1984) La Coopération française dans la géographie bresilienne. In: Cardoso LC, Martinére G (eds) France–Bresil, vingt ans de cooperation. IHEAL/PUG, Paris/Grenoble

Chapter 2
Antecedents: French Geography in Brazil

Abstract This chapter focuses on French geography starting from its tendencies and approaches. It identifies the diffusion of its lines of research through the world and mainly in Brazil, which permits comparative analyses to be established and the identification of differences in the political and cultural contexts of France and Brazil, in the attempt to clarify some approaches regarded as ambiguous. French geography sought to create centers of formation and research, based on the work of a few geographers, the preferential option being Lablachean geography, and using its institutionalization to its own advantage. These centers, at the same time that they were able to create the image of the world in France and of France in the world, became bibliographical references with an enormous capacity for publication and cultural diffusion, often being concerned with explaining a world which was quickly being transformed.

Keywords Tendencies · Human geography · Geographical conception · Cultural diffusion

French geography has always demonstrated much vigor. However, its affirmation stems from the international recognition of the theses formulated by Vidal de La Blache (1845/1918). The most famous French geographer of all times was professor of the Normal Superior School of Paris from 1878 onwards. From this moment, he significantly increased the reach of his action when he occupied the chair of geography at the University of Sorbonne influencing a large number of disciples in France and overseas, especially in the United States of America. La Blache is considered the founder of the French school of geography. It is important to emphasize his main book, *Tableau de Géographie de la France,* which became compulsory reading among French intellectuals in his time. He was also responsible for founding the *Annales de Géographie* journal.

Historically, it can be said that French geography established itself with La Blache being widely applied during the colonial action exercised by France for many years. About this term, Manoel Correia de Andrade with a deep knowledge of French geography affirmed:

© The Author(s) 2016
J.B. da Silva, *French-Brazilian Geography*, SpringerBriefs
in Latin American Studies, DOI 10.1007/978-3-319-31023-7_2

At the start of the twentieth century, France had the second largest empire on the surface of the Earth, which naturally required confusing colonial policies with the humanistic interests of taking civilization to uncultured people who were capable of being educated and absorbed by western civilization, rather than preaching a policy of extermination or conquering these so-called inferior people. (Andrade)[1]

Tropical geography, subsequently developed by French geographers, partly replaced the country's action in that field of science, after the process of decolonialization. The dismantling of the colonial empire did not impede the presence of French geographers alongside the new states being formed in Africa, Asia, and Central America. The main proponent of Colonial geography was Pierre Gourou, of the College de France and the University of Brussels, author of the book Les *pays tropicaux* edited by the PUF in 1947. Another prominent name in Tropical geography is Guy Lasserre, who said the following in an interview given to the author in January 1993:

I am fundamentally a tropical geographer and I created the Center for Studies in Tropical geography (CEGET) in Bordeaux under the auspices of the CNRS.

It could be said that Third World geography is a mixture of colonial geography with tropical geography. However, in fact Third World geography ends up rejecting both colonial and tropical geography which was the cause of the main disagreements of that time and which led to the end of Ceget in Bordeaux. The concrete fact is that today's reality is a globalized world where the market dictates the rules. In this context, French geography adjusted itself to this new global configuration, radiating its influences, trying to adapt to this new reality. In turn, Brazilian geography became known almost eminently for following the French line.

Brazilian geography insists in its self-affirmation. In its attempts to build its own identity in this scientific field, Brazilian geography has faced various types of problems: the excessive number of professional training courses designed for teaching and technical qualifications, issues linked to the definition of the competencies of professors and technical staff, and the weakness of this training regarding the reduced number of professionals involved in the debate to define the paths for geographical science. The discussion about geography as a school discipline, scientific knowledge, and a set of technical–operational instruments has lasted some years. It is possible to list other questions linked to scientific production, dissemination, types of formulations, the role of geography as a science in the national context, corporative struggles in the effort to add value to this field of knowledge, etc.

The concrete fact is that in seeking to understand its origins, its paths or detours, Brazilian geography faced collisions connected to *scientific–cultural loans* or *cultural diffusion* that had been officially installed in the country from 1934 with the creation of the geography course at USP. The contribution of the schools was rarely discussed. It is known that Brazilian geography experienced other influences, such

---

[1]Andrade (1982, p. 184).

as the American and German schools. Geographers of various nationalities and other researchers with work of a geographical nature chose Brazil as the subject of their questions, theoretical formulations, searches, and explanations. Today, the influence exercised by Spanish and Portuguese geographers is strong, a consequence of various partnerships and agreements established between Brazil and these two countries.

It has become common to assert that Brazilian geography is the fruit of French geography. However, the extent of those links is unclear, which is why this study verifies how these relationships were set up.

Brazilian geography should check the designs that defined themselves in the country, as to the centers that established greater or lesser contact with French geography. At the same time, it is necessary to identify the French geographers who choose Brazil as a field of work and research and use the information to verify the level at which these relationships were sustained.

In the recorded cases of Brazilians who were invited to France through technical cooperation programs to take up teaching or research posts, it is implicit that this relationship was not a situation where theoretical–methodological formulations were only transferred from French geography. On the contrary, if this fact is recorded with a certain frequency, it may reveal that Brazil has or would have the capacity to establish what might be called a Brazilian school of geography.

The strong presence of French geography in Brazil is considered by some theorists as one of the obstacles preventing this branch of science from having a different outcome in view of the fact that French geography is regarded as traditional in its general approach. However, it should be taken into account that over time French geography has been able to advance in developing new concepts and applications. Regardless of these advances, there are indications that in the Brazilian case there was the dominance of certain groups crystallized in France, so that some innovative geographical authors were belatedly acknowledged or remain in relative anonymity in the country. In contrast, there were cases that revealed a certain exclusivity or even control in certain branches of human geography, which ended up being interpreted as synonymous of the field. The presence of French geography in Brazil has not always followed a smooth course. There are records of cases of rifts, when certain sectors of the production and dissemination of Brazilian geography, broke with the French school searching for new reference points. The best known case is the IBGE, Brazilian geography's official body, which in the 1970s decided to be guided by theoretical geography, with a quantitative base, of Anglo-Saxon origin.

The effervescence of the 1980s suffered a certain breakdown in the transition to the new decade, with new facts that redrew the world and significantly altered its appearance, such as the fall of the Berlin Wall, the dismantling of the USSR, the emergence of the post-communist bloc of countries, the intensification of nationalist struggles, and the configuration of new power blocs etc....

The study of French geography, its schools and trends, the dissemination of its research around the world and mainly in Brazil allows, especially, the establishment of comparative analyses. It allows an understanding of French and Brazilian

approaches to issues related to the urbanization process, metropolization, the formation of peripheries, social movements, citizenship, everyday life, the imaginary, representation, networks, flows, space, and territory etc. It also allows the exploration of the differences in the political and cultural contexts of France and Brazil in an attempt to clarify some approaches perceived as ambiguous. The bibliographical review and other ongoing research offer the opportunity of a partial perception of the state of the art in geography and especially French human geography. This opportunity has led to some questioning of French geography, with respect to its principles and essentials and its relations with Brazil.

The drafting of the text as a whole evokes a distinct picture from that lived by French geography in the previous century, when it would create theories and models which would be immediately incorporated and applied in countries which were under the rule of French hegemony. The contextual difference lies in observations and questions about the international scope of French geography in present times. The advent of the Internet with its online dissemination of information regarding the new rules imposed by science concerning the conception, the elaboration, and the publication of research outcomes in indexed, internationally recognized periodicals. These changes have shaken the international prestige of many national and even regional schools, demanding the incorporation of new procedures in line with those great centers recognized for spreading innovations. At the same time, the editorial market has to adjust to the requirements of this new period. Virtual bookstores encourage the divulgation and distribution of scientific production. It is in this context that the question is raised concerning whether French geography has been able to follow these changes and if it has maintained the same weight and strength as in previous periods. Geography and geographers who previously occupied a prominent position in those forms regarded as conventional are going through a period of "ostracism" when compared to the new professionals. By this we mean those who edit e-books, do teleconferences, have Internet profiles and circulate the results of their research rapidly. It is evident that geography is not as expressive as it once was. There are several prestigious geographers; however, this prestige does not appear to give prominence to geography, to place it to the fore. On the contrary, the geographers with more prestige are often confused with professionals from other areas of knowledge; thus, there is no resonance capable of giving value and prestige to geography. Of those professionals who are achieving a certain prestige, respectability, and renown, few are known in Brazil. One example is Marcel Roncayolo, the Director of Studies of the School of Higher Studies in Social Sciences, who is the author of various books and articles in the area of urban geography and Urbanism that are widely disseminated in technical and scientific environments in France. Augustine Berque is also in this group, he is recognized for several works on Japan; he is also a Director of Studies of the School of Higher Studies in Social Sciences. There are some others who could be mentioned, but certainly these two examples are sufficient to support our argument. It is worth emphasizing that these geographers almost all work in institutions with a multidisciplinary approach.

Geography has achieved great prestige in the French academic and scientific environment and constitutes a major branch of its foreign policy. This prestige endures, it is unknown if this is founded on the quality of their current formulations or is the legacy of what their geography once represented. Based on this, a myth has been constructed around its quality, dynamism, and application.

In its history, French geography has had years of glory; it based itself on the work of a few geographers and for a while opted for Lablachean geography and through its institutionalization created centers for training and research. At the same time that these centers served to create the image of the World in France and the image of France in the World, they formed a bibliography with an enormous capacity for cultural dissemination and diffusion, which at times focused on explaining a world under transformation.

From descriptions of Colonial geography to the interpretation of France from its regions that became its biggest interpretation, French geography turned to new themes, although these were sometimes treated with old methodologies. Africa was one of the preferential spaces of French geography. Themes like the Third World, Underdevelopment, the Tropical World, and the *amenagement* of territory entered the scene, supporting some geographers and guaranteeing prestige and privileges to geographical science.

Reference

Andrade MC (1982) O PensamentoGeográfico e a Realidade Brasileira. In: NovosRumos da Geografia Brasileira, Santos, Milton, (org.) São Paulo, Hucitec

Chapter 3
The Hegemony of French Geography in Brazil

Abstract This chapter focuses its discussion on the construction of the hegemony of French geography in Brazil and the importance attached to the performance of Pierre Monbeig. It reveals that a consultation carried out with three different authors, one French and two Brazilians, demonstrated a convergence of opinions, although each one emphasized different aspects. It considers the context of Brazil in the 1930s as being a moment with the lack of qualified personnel, meaning that the country had to open its doors to foreigners. Historically, travelers, naturalist, artists, and others would register their impressions concerning the reality of the areas that they visited, focusing on the aspects relating to scenery, characteristic types, spatial arrangement, types of construction, etc. These texts become a priceless documentation for the recovery of the Brazilian spatial dynamic in its beginnings.

Keywords Hegemony · Convergence · Spatial arrangement · Spatial dynamic

The recognition of the excellence of French geography was manifested in Brazil in the choice of France as the best place to offer professional development and qualification. This choice remains in many areas. In the case of geography, there is now a period of retraction due to the setting up of masters and doctoral courses in various Brazilian cities. The appreciation and admiration that Brazilians had for the quality of French geography lasted over half a century, and it is not important to know whether it was elaborated inside or outside France. French professors, especially theses supervisors and laboratory coordinators, were invited to Brazil to minister courses, give lectures, and accompany fieldwork or advise research groups and ministerial teams. Despite this myth and French geography's sensitive reaction in the face of German geography, it is still doubtful whether today it would be entitled to maintain the comfortable position it held former years. Some claim that it is losing its position. In fact, another issue is at stake and the debate should focus on whether Brazil has elected other reference points and whether its scientific production is capable of meeting its own demands and needs. Regardless of the issue of a *myth*, the tradition of establishing scientific and cultural ties between France and Brazil is strong. In this context, geography has taken on an important role and it is not rare to find geographers who exercised strong importance in the Brazilian geographic community.

The arrival of the first French scholars in 1934 is linked to the names of two geographers: Pierre Deffontaines and Pierre Monbeig. The presence of these masters, especially, favored the development of a geographic culture along the French academic and organizational molds, which resulted in academic chairs being established, according to the model that prevailed in France, and also the creation of the Association of Brazilian Geographers (AGB) along the lines of the French organization. Monbeig lived in São Paulo spending eleven years in total in Brazil.

For the purpose of the discussion regarding the construction of the hegemony of French geography in Brazil and the influence represented by Monbeig's role, three authors were consulted: one French and two Brazilians. There was a convergence in their views, although individually each one focuses on different aspects. In this sense, according to Droulers, a geographer at IHEAL and a former student of Prof. Monbeig:

> Pierre Monbeig, throughout his prolonged stay (1935-1946) at the University of São Paulo participated in the particular moment of the introduction of the new geography to Brazil. This means, not on only in the elaboration of university courses but also in the definition and creation of the fields of research.[1]

The text emphasizes Monbeig's entrepreneurial side, citing him as responsible for the introduction of new geography to Brazil.

Manoel Correia de Andrade gave his opinion of the French master by emphasizing his contribution to Brazil through the consolidation of its academic life, which had a significant weight in structuring a Brazilian geographical culture with a French source, as follows:

> His actions at the University of São Paulo were of the greatest importance because he structured and consolidated the chair of geography, subsequently changed to Human geography, and contributed towards the creation of the chairs in Physical geography, which was first held by João Dias da Silveira and of Brazilian geography which was occupied by AroldoAzevedo. The main concern of his work was to analyze problems, reflecting on their causes, the landscapes and problems that they generate and the solutions that could be suggested to solve them.[2]

Maria Isaura Pereira de Queiroz explores another aspect of Monbeig's prolonged stay and his work in Brazil and points out the negative side of few people having had access to the master's high quality works. She states:

> Pierre Monbeig left a large number of studies about Brazil scattered in numerous journals, on both sides of the Atlantic; these precious works only reached a restricted circle of colleagues and students, that is in the scientific community he belonged to. To my knowledge, he did not reach the desks of the leaders of the country or the politicians who aimed to fill important roles in national life. However, these works would have given them

---

[1]Droulers (1991, p. 95).
[2]Andrade (1991, pp. 53–54).

an adequate and profound knowledge of the social reality of the country, its mechanisms and the transformations taking place.[3]

The recognition afforded to Monbeig by these three authors is evident, which reinforces the influence of French geography in the country and the conditions in which its foundations were created. The testimony of Isaura Maria P. de Queiroz is revealing and has a critical and condemnatory character, at the same time that it records a certain disregard for all the knowledge produced at a time when the nation had great need to know its own reality.

The country was growing, accompanying the transfer of population from rural regions to the cities, while at the same time, it was moving inward, following the expansion of coffee production. This Brazilian spatial reality instigates investigation, research, the search, so that faced with the preeminent research needs geographers could not remain immune to this appeal. In response, there was a quantitative and qualitative leap of geography in the country.

A multitude of facts and situations make demands and impose challenges upon the geographer. Of these, the greatest is to respond to social discomfort and demands. Answers need to be supported by new knowledge, which meant the need to build in Brazil a theoretical and methodological framework capable of grasping that reality for the purpose of analysis.

The country needed to deepen its self-knowledge, make a reconnaissance of its territoriality, and establish its cartographic bases in line with the fieldwork taking place. The recently graduated geographer had the task of taming the country in all directions, showing a nation's unfamiliar faces. Brazil was an agricultural giant, with an agricultural export-based economy. The country was in an accelerated process of industrialization and urbanization. These were changeable spaces which revealed differentiated images of the country and demanded qualifications, acuity, and technical rigor. The country was urbanizing fast.

The problems faced when the first professionals were formed in the country became stronger. Geography became a compulsory school discipline. In the scientific field, IBGE increased its staff, carried out research. In the recently created departments, students on geography and history courses (these was considered one single area) did their fieldwork, confirming in concrete reality what was studied and discussed in classrooms.

The latest census data (2010) confirm a continual and agglomerated behavior of the population with practically 85 % of the total concentrated in cities, with a substantial part of this percentage living in large cities. Not only geography but all of science and society are searching for analytical elements to elucidate the country's urban complexity. The progressive increase of population and the advent of a totally new urban reality culminated with the emergence of until then unknown

---

[3]Queiroz, MariaIsaura P. "La Recherche Geographique au Bresil" In: Pierre Monbeig, Un Geographe Pionnier, Op. Cit. p. 64 (T.A.).

social problems. The same is the case for agriculture, with the circulation of persons, products, and capital, leading to new transport needs. The spatial changes were visible in the face of the new logic of territorial organization. The hegemony of road transport substantially alters the organization of space in the country. The older valley cities cut by railroads give way to new cities on plateaus with the implementation of highways and the growth of the automobile industry. Coastal commerce is rearranged as a function of the strengthening of the capitals at the expense of the older ports served by coastal navigation. The relocation of the federal capital and the consequent dominion of the *cerrado* areas of the center-west resulted in structural changes in Brazilian space. The country was rapidly changing its image. Its insertion in the international context favors the emergence of new themes such as the "Amazonian Question" and the "North-eastern Question" in the context of the "National Question." These and other themes instigated and challenged geographers to interpret and discover. These analyses were carried out on different scales. The daily construction of this new reality resulted in the development of a more dependent Brazilian geography, no doubt, at first, on foreign references, especially French. The recognition of the quality of their reflection was obvious. They brought in their formation the concept of applied science. Aziz Ab'Saber, in his contribution, reveals the dimension of the applied nature of the research among French scholars. He affirms:

> Monbeig invested against the abuse of the expression *applied science* in which each group of specialists seeks an application for their area of knowledge without taking into account the strong interactions necessary for application for interaction to take place. In a congress of French scientists Pierre Monbeig defends the idea that there are applications of sciences and not the sole application of one science. On this occasion, the intuitive master that was within him, tried to guide the thinking of his colleagues to the field of interdisciplinary cooperation which is indispensable for the tasks of applying science to the different interests of society and social and economic development. Little did Monbeig know that he was provoking his vain colleagues, each intending to find isolated applications in the scientific field they dedicated themselves to (Ab'Saber 1994).[4]

Brazilian geographers have travelled a long way. Associated with Brazil's dynamics of socio-spatial reality, the numerical disadvantage of a small number of professional geographers to cover a vast country imposed serious obstacles. The lack of qualified personnel opened the country's doors to foreigners. Historically, travelers, naturalists, artists, and others recorded their impressions of the reality of the areas they visited, focusing on the aspects of the landscape, characteristic types, spatial arrangement, and the type of construction. These texts are an essential documentation of the recovery of Brazilian spatial dynamics in its infancy.

---

[4]Ab'Saber (1994, p. 232).

References

Ab'Saber A (1994) Estudos Avançados (8–22)

Andrade MC (1991) Pierre Monbeig e 0 Brasil. In: Théry H, Droulers M (eds) Pierre Monbeig, Un Geographe Pionnier. IHEAL, Paris

Droulers M (1991) Le Developpement de la Geographie Bresilienne. In: Pierre Monbeig, Un Geographe Pionnier. IHEAL, Paris

Chapter 4
Periodization: A Suggestion

Abstract This chapter privileges the discussion of the periods of hegemony, of distancing, of exclusivity, and of new alliances and ruptures that are identified in the relations maintained between French geography and Brazil. It emphasizes the creation of the first university level geography courses in Brazil and the encouragement of the formation of qualified personnel, as well as an affirmation of this field of knowledge, outlining the institutionalization of geography in the country. It recognizes that foreign masters commanded the process of formation of Brazilian geography, providing, in our eyes, a completely new analytical reference point, thus amplifying the systematization of knowledge about the country. From this point onward, France would create roots in the country. A period of strong links begins, leading to various geographers, as well as other professionals, coming to Brazil. This flow becomes diluted as Brazilian universities develop a greater structure establishing laboratories, groups, and research nuclei and institutionalize a capacity building plan and the preparation of specialized personnel to replace foreign guests.

Keywords Exclusivities · Analytical reference point · Alliances · Institutionalization

Any proposal of periodization risks reflects the subjective values of its proponent. Although this is, to a certain point, logical, the different stages discerned in the periodization proposed are nonetheless linked to facts and or events which in themselves are significant enough to be considered marks or boundary points in many possible diverse thematic approaches, independent of any objectives, interests, and methodologies. In the specific case of this analysis of the relations between France and Brazil mediated by geography, the influences, reciprocal exchanges, and the formation of an autonomous school of geography in Brazil are discussed. It is necessary to pursue this path methodologically, seeking its verticality, without forgetting how each fact/event/phase/stage unfolded so as to avoid breaking the

© The Author(s) 2016
J.B. da Silva, *French-Brazilian Geography*, SpringerBriefs
in Latin American Studies, DOI 10.1007/978-3-319-31023-7_4

totality. In this way, the horizontal dimensions are obtained creating an overall picture which indicates a certain direction, way, or path without neglecting the nuances.

In terms of this work, 1934 stands out as a privileged year, for it is the date of the first Brazilian higher education course in geography, marking the beginning of academic geography, scientific geography. The appearance of this more academic geography does not mean that there was no geographical approach in the country, taking into consideration the geographical compendiums and other works of Delgado de Carvalho (Paris, 1884/1990).[1]

For this reason, this periodization proposal favors 1934 treating this date/mark as the beginning of the stage called Approximations. This period presupposes the discussion concerning the setting up of the first higher education geography courses in the country. It marks the beginning of a significant French influence with the presence of French teachers who created geography schools and left a considerable mark on the institutions through which they passed as well as impressing a French "way" of doing geographical science.

The subsequent period, intimately linked to its predecessor, focuses on the discussion concerning the construction of the hegemony of French geography in Brazil. The UGI Congress (International Geography Union) which took place in Brazil in 1956 becomes of utmost importance as a kind of coronation of those events which had begun twenty-two years before. The congress constituted, unavoidably, a fundamental milestone in understanding this path. The presence of French scholars became more intense in the aftermath of the UGI congress. There were many visits from great academics, some lasting longer than others, and the formulation of new theories taking into account Brazil's territorial space and peculiarities.

The hiring of Michel Rochefort in 1961 by IBGE's CNG is the beginning of a phase of alliances between both countries, marked by the arrival of famous French teachers.[2]

---

[1]He taught at the Colégio Pedro II, in Rio de Janeiro. He published various books among which I emphasize the following: *O Brasil Meridional,* in 1910; *Geografia do Brasil,* in 1913; *Meteorologia do Brasil,* in 1916.

[2]Emeritus Professor of the Université de Paris I and President of the Administrative Council of the French Urbanism Institute Rochefort dedicated himself to urban geography and the discussion of socio-spatial inequalities, particularly in developing countries, especially Brazil. Among his works the following stand out: *Dynamiqued'espacefrançais et aménagement du territoire*of 1995 and *Ledéfiurbaindans les pays du Sud,* of 2000. He maintained close connections with different management and development organisms in France and Brazil and with Bernard Kayser the famous professor of the University of Toulouse in the area of agrarian geography, author of various books and articles published in specialized journals. He was a geographer hired by DATAR by the Ministère de l'Agriculture and was president of the European Rural University. He was also a member of CIEU (CentreInterdisciplinaired'ÉtudesUrbaines).

Among the many who were here Pierre George stands out, as he exercised an enormous influence on Brazilian geography. He became an almost obligatory bibliographical reference point in geography courses. In the discussions which focus on the relation between French and Brazilian geography, Pierre George can be treated as an example of a situation of permanence; this is treated under the item Exclusivities which analyzes the fact that independent of technical methodological options, Pierre George's status remained calm and unaltered as he was able to edit his books in Portuguese through DIFEL—European Diffusion of Books, of Sao Paulo, under the auspices of the Presses Universitaires de France, of Paris and by the Fundo de Cultura Publisher. He had a captive audience. The stability achieved by Pierre George does not indicate in anyway that the academic relationship with French scholars continued to be calm.

In an interview granted to this author in November of 1992, Pierre George expressed himself as follows, "… the relationship between my Brazilian friends and colleagues and myself were distended because, at the same time, Brazilian geography was attracted by other correspondents."

The introduction of the new geography of Anglo-Saxon inspiration in Brazil took place in a period of distancing with French geography. Until then, Brazilian geography was treated in a monolithic fashion, as if there was not the existence of groups with distinct preferences and profiles in the various urban centers in the country where there were graduate courses and some postgraduate courses. There is evidence that already at this time there were signs of autonomy being manifest in groups with different lines of understanding, application, and analysis of geographical knowledge.

In Ruptures, we privilege a double rupture. Double in the sense of a break with the practice of geography in Brazil, dominated by the use of the quantitative theoretical models of the new geography, and at the same time a break with what is called traditional French geography. In this phase, the work of Yves Lacoste receives special treatment due to the importance it achieved in this period.

In this way it is possible, in terms of a bloc treatment, to develop a periodization which characterizes this era. This does not mean that this ends the research. Other chapters will be dedicated to specific subjects, focusing on other aspects of the relationship between the two countries. For this purpose, a chapter is dedicated to those geographers who chose Brazil for their research, the direction of their theses, etc., so as to emphasize the arrival of those geography professionals who followed the trail opened from 1934. Another chapter departs from this periodization as it discusses Brazilian academic production in France, in postgraduate programs carried out in that country. To complete the analysis, there is a short chapter to deal with "Brazil in France and how Brazil is viewed by the French," based on rich and interesting reports which allow an approximate understanding of the image provoked by Brazil.

4.1 Approximations: Geography Courses are Established in Brazil

Historically, it can be affirmed that French geography was established in Brazil through the first compendiums produced by Brazilians, dominated by the content of French geography.

The creation of higher education geography courses in Brazil encouraged the formation of a qualified personnel and the positive affirmation of this area of knowledge, leading to the institutionalization of geography in the country. Foreign teachers commanded the process providing a completely new analytic reference from a Brazilian perspective. In this way, the scope of systematic knowledge about the country widened.

Aroldo de Azevedo, a famous lecturer at the University of Sao Paulo, in a brilliant opening speech to the First Brazilian Geographers Congress, held in Ribeirão Preto, in July of 1954, as President of the AGB and of the executive commission of the congress, described the state of teaching of Brazilian geography in terms of production and method as follows:

> The methods which characterize modern geography arrived late in our country; or to be more specific, became widely known a short time ago.[3]

Further on, this great teacher who popularized and gave respectability to geography with his famous didactic books is categorical:

> I do not ignore the fact that, in the first three decades of this century, educated Brazilians demonstrated, through their production, understanding of the modern directions of geography. Euclides da Cunha, while he cannot be considered a geographer, was up to date with the ideas of Ratzel and William Davis, for example. Isolated in his distant Maranhão Raimundo Lopes, during the second decade wrote a book *O Torrão Maranhense*, in which one senses the presence of a true geographer and in which one finds quotations from Emmanuel De Martonne's *Traite de Geographie Physique*. In the same vein are the works of EveraldoBeckheuser, Fernando de Raja Gabaglia and SylvioFrois Abreu, published before 1930.[4]

Aroldo de Azevedo could not avoid paying homage to the great innovator and creator of Brazilian geography, Delgado de Carvalho with the following words:

> However, during this period, the central figure, the great personality of Brazilian geography was, without doubt, the illustrious Carlos Delgado de Carvalho. Every time I read his works, written during this time, my admiration increases for this great Brazilian, already considered by all *the pioneer of modern geography in Brazil*. He already appears as a contemporary geographer when he makes his debut in 1910, publishing his work *Le Bresil Meridional*. He emerges safer in his orientation, without exaggeration a true revolutionary in the routine lived by Brazilian geography when he gave his noteworthy *Geografia do Brasil* to the public in 1913. He displays himself as a true man of science when he writes *Meteorologie du Bresil*, published in 1917. He is undoubtedly a master when, in the third

[3]Azevedo (1956, p. 22).
[4]Azevedo Op. Cit pp. 22–23.

decade of the century, he writes that series of conferences, collected under the general title *Fisiografia do Brasil.*[5]

The French were the most notable of the foreigners, as they stayed and created schools. They were not only geographers for when the first higher education courses were created in Brazil in the 1930s, researchers from various fields of knowledge who would later become known internationally spent part of their academic life in Brazil.[6]

4.1.1 The Phase of Construction of the Hegemony

From this point onward, France would establish roots in Brazil. A period of strong linkages was begun, which occasioned the arrival of various geographers, as well as other French professionals in Brazil. At the University of São Paulo (USP): P. Deffontaines, P. Monbeig, Emile Coornaert, Fernand Braudel, Claude Levi-Strauss, Paul Arrousse-Bastide, Etienne Borne, JealMangiie, Robert Garric, Pierre Hourcade, François Perroux, Rene Coutin, Pierre Fromont—At the University of the Federal District (Rio de Janeiro), Henri Houser, Gaston Leduc, Maurice Bye—In Porto Alegre, Jacques Lambert.[7]

This flow becomes diluted as Brazilian universities improve their structures with the installation of laboratories, groups, and research centers and institutionalize a plan of capacity building and training specialized personnel to substitute the foreign guests. In the context of the policies developed in the postwar period, CAPES is created in 1951 as the Personal Development Campaign of Personnel in Higher Education. This agency, which had the role of assuring the existence of specialized personnel in the quantity and quality necessary to meet the needs of private and public enterprises, acting alongside universities and higher education institutions was linked to the Ministry of Education and Health.

From the employment of visiting geographers as lecturers in Brazilian universities, providing a general analysis of the Brazilian geographical situation, one arrives at the UGI Congress (International geography Union).

4.1.2 The UGI Congress

The UGI Congress was an event with a great historical value. Nice Lecocq Muller described this event as follows, "the planning came before the founding of new cities with rigorous criteria of locations which allowed the access of the rural

---

[5]Azevedo Op. Cit p. 23.

[6]For a more detailed treatment, see Monteiro (1980).

[7]Chonchol and Martiniere (1985, p. 90).

population."[8] With regard to the UGI Congress, she was emphatic in affirming: "…
beyond stimulating the renewal of points of views and methods through contact
with foreign specialists, it stimulated a series of urban studies."[9]

The Congress marked the beginning of contact with a more specialized geography. The event was essential in establishing new relations. From it the flow of
Brazilians toward France seeking professional development became more intense,
as did the interest in specialization.

Orlando Valverde, with his profound knowledge of the history of Brazilian
geography, described this congress, emphasizing its exceptional nature and
importance, in the following terms:

> An exceptional recent episode of cultural relations in terms of geography was the opportunity offered by the XVIII International Geography Congress which took place in Rio de
> Janeiro from August 3-18, 1965. For the first time the International Geography Union
> (UGI) sponsored a global colloquium below the tropics and in the southern hemisphere.
> The French delegation was not the most numerous but was without doubt one of the most
> brilliant … Maximilien Sorre, Pierre George, Jean Drech, Jean Tricart, Pierre Birot, Andre
> Cailleux, Jacqueline Beaujeu-Garnier, Michel Rochefort, Bernard Kayser, P. Deffontaines,
> P. Monbeig,… B. Kayser revealed a new field of ideas of great social and economic interest
> for Brazil: labor geography … M. Rochefort attracted much interest for the course he gave
> at the National Geography Council, where he introduced the notions of service geography
> and the concepts of urban centers and networks.[10]

From that moment on, geography is established in an ascendant trajectory in
which the IBGE starts to have a noteworthy role. The intensification of the
exchanges led to contact with a specific geographical literature which achieved
great production in the 1950s and 1960s.

This significant production of the basic foundational texts of French geography
contributed to fill a specific methodological and theoretical gap in geographical
science in the treatment of themes demanded by the new reality. To verify if there
was a difference in the treatment of distinct situations concerning the social–spatial
realities between France and Brazil, various French geographers were interviewed,
especially those who maintained scientific work relationships with Brazil.

In reply, Jean Labasse, renowned geographer of the University of Lyon, gave the
following report:

> My work in Brazil taught me much, especially about the functioning of urban networks.
> I also owe to my time in Brazil a certain way of appreciating large spaces, which I did not
> obtain in other places as my periods in Canada were limited to the eastern provinces and my
> trip to the USSR was very brief.[11]

---

[8]Nice Lecocq Muller was one of the most influential geographers in the country. One of
Monbeig's students she specialized in urban geography. She dedicated herself to urban and
regional planning. She is the author of the renown book *O FatoUrbanonaBacia do Rio Paraíba*,
de 1969.

[9]Muller (1968, p. 16).

[10]Valverde (1989, p. 83).

[11]Interview carried out by the author in December 1992.

Labasse emphasizes the difference in scale, a fundamental fact in the process of differentiation. In this same vein, the interview with the geographer Bernard Kayser provided an equally emphatic answer.

If I had not 'worked' in Brazil, I would not have developed this intuitive doctrine towards the Third World which allowed me to develop courses and write articles many times contrary to common opinions and analyses.[12] Kayser makes the differences in socio-spatial context evident. He attributes the changes in his concepts about space to his time in Brazil.

4.1.3 Distancing: The "New Geography"… and France Became Regarded by Leaders in Brazilian Geography as a Reference Point of the Old, the Past

At the start of the 1960s, Brazilian geography was fermented by the search for its own direction and the need to provide answers to the rapid transformations of the dynamic of Brazilian territory and society. At the same time, it became attached, in official terms, to an orientation which was very different from that which existed after the UGI Congress. A new country was being drawn, and the repressive state resulting from the 1964 coup soon sought to use geography not only to carry out war but to also annihilate where ideas and resistance might be formed. Where a small level of liberty remained it was still monitored. The state became militarized and organized sophisticated forms of control. Geography was being prepared to give its answer. The official response became embedded with status. A geography elaborated on mathematical terms was, in the eyes of those technicians anxious for a more exact and scientific geography, based on formulas and models and thus able to gain greater application and recognition. Many Brazilian geography executives achieved a stronger position after the coup. The flow between Rio de Janeiro, Brasilia, and Washington became more intense with society being the loser and the state the winner in a context which became evermore repressive and excluding.

Monteiro (1980) proposes 1968 as a dividing mark for what he considered the "official proclamation" of the addition of new practices to geographical analyses.[13]

Following this line of thinking, the introduction by Faissol (1978) to his book *Urbanização e Regionalização* which was published by the IBGE uses language which was until then incomprehensible to the majority of Brazilian geographers to introduce the article by Gunnar Olsson (note that ten years had passed since the introduction of theoretical geography in Brazil).

---

[12]Idem.

[13]Monteiro (1980).

...following Michael Dacey's methodological line using point distribution models that follow different probability laws (the most frequent are Poisson, the distribution families of the contagion type such as the Negative Binomial).[14]

4.1.4 Exclusivities: Pierre George

Pierre George became a kind of national consensus with regard to geography. His works edited in Brazil have been used for many years in several of our academic courses. As a researcher, he produced a considerable volume of texts including books, essays, research reports, articles in specialized journals, conferences.

These days the presence of Pierre George in academic circles is more tenuous as due to the carelessness or deformation in our pedagogical and didactic practices authors read as classics are regarded by geography university lecturers as outdated. This observation is valid for various authors, including Brazilians, whose vast experience in scientific production remains forgotten on the shelves of libraries or in archives. This exaggerated quest for what is new, while undoubtedly interesting, errs through the loss of the historicity of the process of knowledge production. As an example, one could ask how many students have read Milton Santos' classic book *A Cidade dos PaísesSubdesenvolvidos*? The same can be asked in relation to Manoel Correia de Andrade. Young geography scholars do not read his often cited *A Terra e o Homemno Nordeste*, another classic of Brazilian geography, even though it is obligatory reading for students on other courses. The same could be said about the international *Geografia da Fome,* by Josué de Castro—who remains an illustrious unknown to our students. The analysis of Pierre George's situation is based upon his status as a well-distributed author in Brazil, with a significant part of his work edited in the country by DIFEL.[15]

Speaking about his experiences in research and across the world, in which he includes Brazil, George expresses himself as follows:

> In 1946 the discovery of a 'new world' begins with the first contact with Tunisia, which will be followed with various periods of being a lecturer at the University of Tunis ... Later, the journeys to Morocco guided by F. Joly, Jean Le Coz, and Daniel Noin. However, the main fields of observation were the journeys linked with participation in congresses, colloquiums and periods of teaching in Latin America: Brazil, 1956, 1962, 1968, Chile, 1966, Argentina, 1965, 1969, Venezuela, 1977, Mexico, 1964, 1976, 1978, 1980, 1982, 1984.[16]

---

[14]Faissol (1978, p. 22).

[15]Among the most important we cite: OsgrandesMercados do Mundo, Third edition, 1969; Geografia Industrial do Mundo, fifth edition, 1979; A Geografia do Consumo, second edition, 1971; Geografia Social do Mundo, 1969; Geografia da População, fifth edition, 1978; Geografia da URSS, 1970; GeografiaAgrícola do Mundo, Third edition, 1975; A Ação do Homem, 1971; OsMétodos da Geografia, second edition, 1978; 0 MeioAmbiente, 1973.; População e Povoamento, 1975; Panorama do MundoAtual, sixth edition, 1979; Populaçõesativas, 1979.

[16]George (1990, p. 235).

The current relevance of George can be observed by an analysis of the content of various conferences carried out in Brazil in 1962 at the IBGE in Rio de Janeiro. In the quotation extracted from the text of a conference from 1963 when he dealt with the theme of "Some problems in the geographical study of population," he reveals himself to be an up-to-date geographer. In the cited portion, the renown geographer analyzes the reality of the future technopoles becoming the dominant industrial scenery in developed countries:

> The location only depends on the acceptance and choice by men which is increasingly difficult as a geographical installation. What matters for them, effectively, is being able to make the most of the advantages offered by modern techniques without facing any inconveniences. New types of locations in places which are more pleasant to dwell in and better supplied with services (Grenoble in France) substitute the ancient fixation in the somber carboniferous regions or the sad suburbs.[17]

George had a very accurate knowledge. He gained recognition in France where he was called to direct or participate in planning and management teams. He was a wise and critical geographer. He was a witness of the advances achieved by French geography, developing a severe judgment concerning the excessive level of fragmentation of scientific geography. He is scathing and rigorous when he reveals the different compartments of laboratories and research nuclei which do not communicate together.

Pierre George in his magnificent book, *Le Métier de Géographe*, which celebrates his fifty years of geography, collects a set of articles which reveals his disquiet and anguish concerning the trajectory of geographical science. In *Annales de géographie*,[18] he discusses the "Difficulties and uncertainties of geography." The text reveals a professional who is up to date with what is going on. The text is a forerunner, in which George deals with the interrogation of a category which asks itself—"Is geography really a science in itself, considering what it does with its methods and its diligences"? Trying to answer these questions when faced with the difficulties faced by geography, the author continues with his questions "What remains of geography? A sign above the door through which different corridors lead to the different laboratories of geomorphology, of the study of soils, the study of sediments, of climatology, more or less meteorological, but also agrarian structures, demographics, urbanism, the study of transport where there is a duel between technicians who ignore one another." He criticizes scientism which he clams leads only to a break into different sectors and fragmentation and that these lead to a focus on schematics with a tendency to perfectionism.

His maturity is expressed in the development of a critical conscience when he speaks about the limits of scientific geography when faced with the challenges imposed by the new times.

---

[17]George (1963, p. 31).
[18]George (1976, pp. 48–63).

The moment comes when the geographers' arsenal appears to be unused or inadequate to continue with a movement which appears to be diversified and accelerates to the point that the traditional university system with the production of exhaustive theses appears to be completely unsuitable, incapable of following the rhythm of events.[19]

He demanded a geography with greater coherence, more objectivity. He declared: "A science is defined by its object and not by its methods. Every method which does not adapt to the object is unsuitable."[20]

Among other publications by George about Brazil, one can mention: "Originalite des capitales des pays temperes de l'AmeriqueLatine", *Revista Geografica, InstitutoPanamericano de Geografia e Historia*, Rio de Janeiro, December 1967, pp. 31–42; "Des routes de l'aventure aux pullmans des autoroutes au Bresil", *Transport, Paris*, 1959, pp. 112–118; "AireMetropolitaine, conurbation ou region industrielle, les cas de São Paulo", in *Regionalisation de l'Espace au Bresil*, Bordeaux, Centre d'EtudesGeographiques des regions tropicales et Paris nCNRS, 1971, pp 175–183.

The period known in Brazil as one of theoretical and methodological renewals in geography coincided with a strong distancing of Brazilian geographers in relation to the positions held by Pierre George. Yet, the scale of his work and the number of editions of his various books are evidence of his importance in Brazil.

4.1.5 New Alliances: Rochefort and Kayser: The Urban and the Rural

The reflections of Carlos Augusto Monteiro about the hiring of Rochefort by the IBGE and the number of study, research, and consultancy trips he made to Brazil point to the significance of his presence among us and the support that he gave to the relationship between the two countries in terms of cultural and scientific exchanges. Rochefort is the symbol of a period in which France was much sought after for the supervision of theses and in this he has a leading role.

The influence of Michel Rochefort was not only present in works based on the application of his method. His ideas, his orientation regarding how to face the problem of urban networks would remain registered in other more recent works, independently of the research method adopted.[21]

The full transcript of an interview carried out in December 1992 in Paris is provided to provide a deeper reflection on the impact of Rochefort's arrival on the scientific relationship between Brazil and France:

---

[19]George (1990).

[20]Id. Ibid., p. 81.

[21]Correia (1968, p. 192).

4.1.5.1 Interview with Michel Rochefort

Among the many activities developed by Michel Rochefort in Brazil, the most well-known is that of advisor. However, he also carried out activities linked to teaching. Rochefort affirms that he taught the application of geography for the service of *aménagement,* while teaching for research purposes and solely for lectures only on a few missions.

I was, for example for at Recife University for one year, in 1961-1962. I spent the year teaching the application of geography to regional and urban *aménagement* geography, but applied geography, geography seeking the *aménagement* of the territory.

He continues:

After this I participated in missions for which I was invited essentially for *aménagement* services (1966-1967) in 1966 via CEPAL and in 1967 at IPEA. The director of IBGE at that time was Mrs. Velloso. After this I worked in the CNPU- the National Commission of Urban Policies, and from 1974, with an interruption in 1967-1968 I worked in São Paulo with Maria Adélia in the *aménagement* Ministry or State Planning Secretariat in the middle sized cities project. I worked in Brasilia for the CNPU and each time I would carry out conferences at Universities but I had not gone specifically to teach.

About more recent activities he affirms:

And only a few years ago I did a special mission to teach on the technical cooperation team between France and Brazil (CAPES/COFECUB) in Salvador and Belém. I was in Brasilia for a CNPU Mission. Another time I went to Fortaleza only for a few days to give a few conferences. I was in Brazil, I believe twenty-six (26) times from 1956 onwards. I went twenty six times to Brazil and each time I did conferences in universities, I might say all the Brazilian universities.

About these conferences he declares:

I did conferences in Porto Alegre, Florianópolis, Curitiba, and São Paulo, but also in Presidente Prudente, Recife, Salvador, Belém, Brasilia, Vitória and each time teaching was a secondary function in relation to the main mission which was a technical cooperation and service to the Brazilian *aménagement* to apply geography. And each time I was working for the agreement between France and Brazil. The first agreement, I think, dates from 1948. You know, there was a permanent technical cooperation service between France and Brazil. I think that M. Pebayale was the last French person to go to Brazil through this agreement. It's possible. These were agreements made with academic objectives. Those I participated in were the French foreign affairs' permanent technical and scientific agreements. These agreements with Brazil financed my mission and secondarily, Brazil permitted me to carry out conferences in the universities. With Maria Adélia I began to have responsibilities in the Planning Secretariat (1974) where she was the regional action coordinator of the Secretariat. My university role consisted of receiving students with their research, their theses rather than teaching in Brazilian universities. These were students coming from various universities in the countries and for me, this activity meant more than teaching in Brazil.

Rochefort, with his acumen, adjusts his discourse, adapting himself to Brazilian reality. He affirms that the Brazil of today is far more complex. To deal with the country as a whole is very difficult for a foreigner. Current researchers tend to focus on individual states in the federation or regions. There is evidence of

an increasing focus on the particular, and the deepening of the orientation mechanisms of the new geography, the geography of actors, the analysis of mechanisms; I believe that when one moves from the description of the landscape to the analysis of the mechanisms of the relationship between space and society, anyhow in the geography of actors the particular is studied so as to better perceive these true relations. One moves to a geography of correlation where there is such a landscape and something else, etc. When I changed this part of my activity I went to be a teacher in Togo, at the University of Daome to conclude a thesis under my supervision. Well, I always tried, all the time, even when I was in Brazil, to maintain a cooperative activity with urban research in Africa, specifically in the Maghreb, tropical Africa. I supervised a certain number of theses there, about the Maghreb, especially on the famous thesis of Signolles, in Tunisia, about Tunisian space; I supervised a very interesting thesis which has just been examined which is called "Urban poverty and residential mobility in Marrakech."

Motivated by his professional practice as a research director, the professor continues discussing the theme and content of the thesis, especially concerning the methodology adopted:

It is a new approach which takes the social phenomenon of poverty to understand how one can best explain the mechanisms for the exchange of residences in the urban and suburban space, the social spatial segregation and the mobility of the poor in housing research where they are not only marked by simple segregation, for there is a capacity for mobility within the spaces of poverty. You would say that these spaces of poverty are part of a dynamic because there is, especially in the case of immigrants, a space of "hospitality." These are someone's space of solidarity and only in this way do they manage to escape this history of poverty.

Rochefort gives his opinion about the relationship between France and Brazil in relation to geographical academic activity, discussing the phases of this relationship:

The two countries lived moments of close proximity, moments of apparent rupture and other moments marked by the absence of Brazilians enrolled on French postgraduate courses. Beyond this, it seems that there are different theoretical conceptual approaches, especially in France when the discussion regarding the collapse of Marxism becomes incorporated, differently from what happened in Brazil.

He affirms that this dynamic is extremely logical, that these are two different questions from the perspective of the relationship between French thought and Brazilian geography.

With regard to his activity in other countries, he declares:

In the countries that I collaborated with I perceived that there was a dynamic in which, to begin with, a certain number of concepts and notions are received from a geography which wants to be universal but is also French and therefore has concepts and notions which are marked at the same time by the structure of French thought and the uniqueness of the French space where it was worked out. This is the infancy when one receives concepts and notions from others, and there is, by necessity, a moment in which someone rebels- this is legitimate! This rebellion against these concepts is the moment of seeking autonomy, and to seek autonomy over the planning of means. Finally, Brazilians are closely linked to the study of their country. Brazilian geography received external concepts, but Brazilians applied them to the understanding of their country and it is, in truth, a nation where geography is linked more closely to the idea of the nation. You mentioned that the geographers on the National Geography Council and those of the university were closely linked to a certain social role that they should play as geographers so as to better know their country and make it known. It is a true, respectable necessary nationalism. At a certain moment there is a rebellion against foreign concepts, or better, to understand Brazil it is necessary to find anything which is better adapted to this reality that one wishes to analyze, and find a way of understanding the relationships with Brazilian society. So there is a quest- a specific Brazilian society sees that our geography is an adaptation of the spatial search of this particular structure of Brazilian society to the extent that it is no longer the space of this society. At this time, true Marxism is very important in this search for the relationships between society and space. So, we have a whole generation of Brazilian geographers who are exaggeratedly Marxist and possibly with a hyper tendency to reject anything which is not necessarily Brazilian. I can accept this tendency because I speak Portuguese, it is easier, but it is truly an attitude which I call positive rebellion. There is this movement and we can say, without citing any names, that there are a certain number of Brazilian geographers who critically regard this strand of Brazilian geography, who want to consider themselves independent but who continue to do theses in France and write Marxist texts at a time when in France all of this Marxist structure of thought is being questioned, so that these texts are not well accepted. That is true and it caused some difficulties. I would also add that Brazil faced a difficult period with the military dictatorship which did not facilitate an approximation with France. I was criticized for going to Brazil during the military dictatorship, they asked me, "Oh, well, what are you going to do there… etc.?" And I worked with people who were against the military dictatorship and denounced it and were the moral support for all of us. This is evident and it is true, but afterwards, I do not know why, there was a renewal of French geography and there is a beautiful phase in Brazilian geography and they are capable of approximation, but now these are adult relations. There was the infant phase when Brazilian geography was formed by French teachers, there was an adolescent phase when Brazilian geography rejected its parents and, finally, now there are adult relations when there is a conversation, there is an exchange, but neither has a supremacy over the other. But I think that by now fifteen years have gone by and I repeat this every time I go to Brazil that there are many things in Brazil by Brazilian geographers, more than everything that is usually thought and expected. All that is thought is precisely to disentangle oneself from general concepts as Brazil has become closed within Brazilian geography. An approach that can be undertaken by French geographers is a comparative approach, comparing with another country. Due to my specialization in Third World cities I do make this comparison, but it is different in Africa, it is different in America but the understanding of Brazilian geography has been made infinitely deeper by Brazilian geographers than by anything that the French could do. French geographers do not do more than specific research, as for example, into the informal sector in terms of its relationship with the rest, but Brazilian geography now belongs to Brazilians and it is for this reason that I am displeased to see books about Brazil written by French authors without Brazilian collaboration. I think the time has arrived in which to write about Brazil, if France is going to write a book about Brazilian geography, it is necessary to ask a Brazilian to write it and the French should translate it. We and the Brazilians now have an adult relationship of comparison, so that Brazilian geography deals with Brazil and we will bring the vision of other spaces, and one translates and compares.

When asked about the fact that the country produces a very Brazilian geography and if this would be a problem or simply a condition caused by the complexity of the country with regard to its extension and surface, he answered:

This is something that was created because Brazilians have this nationalist motivation, and Brazilian geographers are often engaged persons who want to do anything for their country. So they are scholars of their country and perceive that this country is so complex that it is sufficient as a study lab, and there are differences between the Amazon and Rio Grande do Sul.

Concluding his discourse, he affirms that

A problem that is yet to be resolved is the relationship between Brazilian geography and the geography of other South American countries.

He continues:

This has not yet been resolved because you find it easier to relate with the Europeans, the North Americans than to your colleagues from South America. This is a problem that can yet be resolved, but lacks time. At the colloquiums, the current congresses what seems to me to be the most interesting is this affirmation of Brazilian geography. It is established as the only one which knows Brazilian problems best, and which brings this knowledge into a general discussion with other geographers, but does not add what should be translated. Now it is true that this should also exist with the other South American countries. I dream of a congress in which geographers from Chile, Peru, Colombia, Argentina and Brazilian carry out egalitarian exchanges on a subject which is of Latin American interest.[22]

When asked if the relationships would be the same between the French and Brazilian, or it would be different between the French and Peruvians or between geographers of other countries and Mexico which does not yet have a recognized geographical school, he answered:

That is the problem. Geography in the other Latin American countries has not known the same evolution as Brazilian geography. There are some countries which have a geography school but there are others that have advantages in terms of urban sociology and do not have the same approach and for this reason you have this difficulty of going to the end. I think that this adult geography of Brazil has the original input from France which created the Brazilian geography school which starts with an advantage and you are much more ease in talking with the Maghreb, with black Africa because there is this sort of starting point. There are better than before, fundamental concepts which are concepts from the French school. This is what makes the dialogue easier, even if each one has evolved, found its own knowledge, its own concept, there is still a base which allows an easier dialogue than that with other countries which have not known the same evolution.

Returning to the issue of the phase of rebellion that occurred in the relationship between the two countries, what Rochefort classifies as infancy and adulthood, I provoked the professor saying that for us Brazilians, especially for myself, there were indicators that there were also changes in France, a kind of internal rebellion

---

[22]This interview was carried out before the EGAL meetings were carried out, a privileged space in Latin American geography.

and that he had a significant role when he worked on the concept of territory *amenagement*, of urban *amenagement*.

Rochefort accepts the provocation and responds:

Certainly in Brazil and in France there are always numerous famillesd'esprit or currents of thought and there are people and there are still older geographers in Brazil who remained in the infantile state. There are still Brazilian geographers who are in adolescence, who are still rebellious towards all foreign collaboration, saying Brazil is for the Brazilians and there are adult Brazilian geographers. In France it is the same. There are geographers who represent the French geography from forty years ago, who have not evolved and there are certainly internal rebellions. I followed the rebellion of the AGB against their mandarins in 1978 in Fortaleza, very closely, but this was an excessively harsh judgment. But there is an effort to clarify a situation which has been left behind. There are people it is necessary to pay attention to. I am, for example, very friendly with Manoel Correia de Andrade, for me he is a very close friend, a great friend and I am very hurt to see Manoel Correia de Andrade rejected by the young, young geographers who are a touch exaggerated. This is harmful, Brazil is always a bit violent, and it is the violence of Brazilian society. And here in France there is certainly renovation, rebellion, but these are always with different nuances and they are always *famillesd'espirit* that differ one from the other, there are different tendencies and one can say this about French geography and one can also say this about Brazilian geography. There are tendencies that I know well on the French side and, the only ones I know on the Brazilian side are those which fragment geography.

I continued with the provocation affirming that in France there are specializations, in Strasbourg, which plays a different role from Bordeaux, Toulouse, from Paris.

Rochefort accepts this and continues his revelatory discourse:

That is true, and it was reaffirmed by personalities. For a long time French geography was dominated by personalities as it was a mandarin system, it was dominated by a few personalities who, at the same time, had the power in the provincial universities. While in Paris there was a crowd of teachers of diverse tendencies, Strasbourg was marked by Tricart, in Bordeaux Laserre marked supervision in Lyon the same happens with Labasse in Lyon. There are many more cases of personalities etc. But that has passed, these were moments. Now the universities are much more varied.

When asked if the RECLUS group and the Maison de Geographie de Montpellier-MGM would be the new French geography of today, he answers with his characteristic calm:

It is not a French geography; it is an orientation, but one orientation in the midst of many others. This does not represent the future of geography for me, it represents a tendency, which is full of many interesting resources, means, techniques, sources of financing, which brings us to the goal of this tendency which is the use of modern techniques, databases, to arrive at a modern cartography, but I think that this is extremely reduced in relation to the geography that interests me, which is the geography of mechanisms, of understanding the interaction between social, economic and spatial mechanisms and which refuses this modern form of finding the way to describe the landscape. But when they claim to find the only conclusion about mechanisms it becomes very dangerous. I worked a lot with territory *amenagement*, and I think that to establish as the privileged axis of European development, what they call the "banana"[23] I think that this causes enormous harm because it is a simplistic view of space. This group has many resources enormous means, and they have a leader, Roger Brunet who is my friend, who has done much, but it is always an orientation, a useful orientation, a new position with modern techniques, a new way of working space and statistical data. This is not for me the whole of geography but it is an interesting aspect of geography which is a new description of space thanks to the new way of treating data; but one can make serious mistakes in the understanding of mechanisms; for this reason it is necessary to make the effort to understand the geography of actors, because they are the actors of the interaction between the economic, the social and the spatial, the domain of geography which interests me is that which I work on and worked on. I do not say that this is not geography, but it is a complementary step in geography, and one helps the other, but one does not have supremacy over the other and both are important and if you want to do territory *amenagement* this is not enough. Description is not enough, it is necessary to understand the mechanisms, because it is the mechanisms that lead to action. For me, at this time, I think that the geography which is most useful for society is the geography which allows one to perceive and change things in the society-space relationship. And to act, to change anything, it is necessary to understand the mechanisms. If one does not understand the mechanisms, nothing gets changed. For this reason I respect the RECLUS group, but for me this is not geography. Claval has a different orientation from mine, but I respect his path enormously, which is to understand the cultural domains. There is a gap, a gap marked by Marxism - not him, but me- but there is a privileged gap of social and economic mechanisms and it is thought that the social is linked to the economic; it is nothing but the expression between the social and the spatial, the social is also cultural and therefore what one wants to understand, it is simply necessary to know not only the social and cultural mechanisms but also how the cultural has an important role in the mechanisms of interaction between society and space. And for this reason Claval's work is so important. The same effort is necessary to understand the interactions between space and society. Claval is a culturalist, and I on the other hand focus on the economical, the social, but the social categories are complementary strands.

My next question was with regard to the singularity of geography traditionally sought in the relations between physical geography and human geography, its two most important branches. Considering the fact that space is social I asked Professor Rochefort whether in this case we were not attributing to physical geography a merely instrumental function. In his answer, Rochefort reveals his profound knowledge of geographical science, a comprehension which provides meaning to the questions linked to the singularity of geography. On this issue he replies that

> For France this is very serious because in our journey we have had the need for a physical knowledge of space, knowledge of the natural domain, a need for more and more knowledge of the environment, but we also have the need for knowledge concerning the current mechanisms of the relationship between society and its natural habitat. In this way it is all of physical geography that will provide us with the data, for it is shaped for the analysis of the environment, the physical aspect of the environment and when one considers the influence of the geography of natural risks it is an indispensable geography. But physical geography does not just want this; it wants to be a geography that explains the natural environment that explains as the background of a picture but also its evolution, the reality of nature and the path taken by natural sciences. In this sense geography is not well

---

[23]The "banana" described in a prospective letter of the RECLUS group refers to a stretch of land following a NW/SE diagonal along the European continent, stretching from England, the London basin region, to the North of Italy, the Milan and Turin region.

equipped, because climatology, geology and meteorology have undoubtedly done this much better. And physical geography either accepts becoming a geography of the physical environment in its relationship with society, which is indispensable and no one argues about this, or it will simply be a physical geography for understanding the natural sciences and understanding nature at that time, geography is lost.

I remind the professor that in Brazil this is a recurring discussion and he continues:

Human geography we know, it is social space; it is social but it is space. But the spatial data is not something which is malleable; it is reality, the interaction between actors. Actors are the physical reality of space.

Few professionals have known Brazil with the level of depth constructed by Rochefort. He has been in Brazil twenty-six times and in this coming and going is able to reflect on the role Brazil has played on his development and intellectual itinerary. In response to this he affirms:

Oh, this was very important! Brazil was for me; I can say, in 1956, the discovery that geography was not reduced to the developed world. Before that I was truly centered on France. Then I started to realize that there was reality and it gave me the spirit, the idea that it was necessary above all to start from the reality of a society if one wanted to understand human geography. Before this I focused on humanizing general concepts, humanize the city, etc. and did not yet see the reality of the city relative to the structure of society. I discovered a society very different in a rural perspective, with the heritage of farmers, I also discovered a very different urban society with the burden of poverty and on this journey I told myself "well, this society has particularities and it is necessary that I relativize my human geography towards the reality of each society." This was a wake-up call for me, and from this awakening, I realized that there was a fundamental difference between developed countries, my apologies, and underdeveloped countries and it was from this that starting from Brazilian knowledge I became interested in what we call the Third World. Now Brazil is a bit more distant from this condition and I started to work in Africa, about Daome, etc. but Brazil was always in my thoughts. This discovery was at the same time a discovery of Brazil and what the Brazilian geographers told me about Brazil and what I did in Brazil in the period from 1956 to 1961. One should not forget that between 1960 and 1961 I spent a year and a half in Brazil. In this year and a half with Manoel Correia de Andrade and Mario Lacerda de Melo I did a *tournee* through the inland regions of the Brazilian northeast and spent two months there and realized that my concepts of "city" and my concepts of "time" fell apart ... and this was fundamental for me, and I became another geographer, different from the one who wrote a thesis about urban organization in France.

I took the opportunity to say that I had much interest in discussing the issue of tropical geography and of colonial geography in the context of French relations with Brazil. I asked concerning the possibility of finding any common point in the two conceptions. Immediately, Rochefort revealed his wide and deep knowledge. He reached the crucial point of the question and gave ample opportunity to establish the relations when he affirmed that

From the perspective of French geography, the geography applied to studies about Brazil, is to a certain extent similar to tropical geography and colonial geography; this from the point of view of the French geographers (laughs). I think that I never did colonial geography and never did tropical geography because for me, both are false concepts. One is a temporal reality for certain countries, for a certain country ... but I referred to the geography of actors. In certain countries there were certain actors so that in a certain period there was colonialism. There is the colonial fact with the actors of the current social-spatial reality. Tropical geography states that the tropical environment is also an actor. They say about the environment, it is true, that geography is in fact similar to the natural physical, tropical environment. But it is still just an element. It is necessary to understand the global relations between the society and space, as in African society. It is true that in African society there is a colonial reality which was much used, there is a tropical reality, but there is also the people, this people who were colonized, who did one thing before and another after. So it is always necessary to do the geography of every society in its relations with space, and for me tropical or colonial are elements in the sense of a globalization it is necessary to pay attention to. I think that in fact, I neither did colonial geography, nor tropical geography. I think, although this is unpleasant for some colleagues, that certain French geographers did much colonial geography in Africa and tropical geography in Brazil and from this they privileged their partial concepts of geography.

The great teacher, with his deep knowledge of Brazilian geography, looks at me, tries a smile and asks "Did I answer?" At this stage of the interview I was enthusiastic with the result achieved. Rochefort, motivated by the subject and affable as always, continued:

In terms of geography, in France there are orientations that I respect deeply, such as the quantitative geography which brings us an infinitely more solid vision of reality. However, in terms of the comprehension of relations it is very fragile because its factorial analysis provides the correlation but not the understanding of the role of each one of the group of actors in the mechanism of interaction between society and space. These mechanisms of interaction provide infinitely more precise analyses than can be made by quantitative geography. And this consists of knowing the actor and knowing the mechanism of power concerning the space of different societies.

With regard to his current research projects being currently developed and the geography of actors, he affirms:

It is there, precisely, that effectively there was a phase in which Marxism gave a very easy appearance because the geography of actors is the spatial translation of class conflict. This took place, one learns something and it is very useful for evolution. This was lost, that is true, and it is necessary to recover a certain number of forces that act in an orderly manner upon space, the urban and, if you want, act with different temporalities. So the geography of actors has already sought to see the temporality of each group of actors, and, in this way, one finds the idea of the heritage of the previous structure, and the geography of actors understands that there are actors whose role is still present, because there is the permanence of what they did in space and there are different temporalities at the same time in the way that they act upon space and the way in which their action upon space becomes permanent. It is difficult to change.

When he defines the profile of each actor and the results of their actions he states "You have to decide who can act upon the space and in this decision you not only have the State but also the whole range of different levels of current public power." Discussing public power he affirms:

In this way it is the public power which is invested with power upon space. At this time, the problem is to know in whose name this power acts upon space and one finds the dilemma of the lack of ideology and one can say that it acts upon space in the service of capitalism, or not, that it acts upon the space in service of the common good or that it acts upon space in terms of a certain number of prejudices and a certain number of models that are being considered. And it is this which is very difficult to reduce into a theory. The action of public power is done through people who have the power and search for a model, it might be an ideological model, it might be a sentimental model, and it might be an interest model. There are some studies, for example, about the role of the mayor in the peri-urbanization around Paris and one can see the mayor's motivation through the analysis that he will make when he has to decide something. You can see what motivations are present in the public actors at different levels. There is a Marxist role of interests in relation to capitalism. I think there is a joining of models and also of concepts of general interest, of private interest, etc. You can see the directors of a private company for whom space is a source of income and you can watch all the strategic territorial games by these directors, who now have a great power of mobilization because they can decide the location of their activities. And so, what exists is a territorial strategy of these private company directors, and there is the role of the inhabitants, who can become citizens in the full sense of the word if they become organized in associations, in committees to act upon the space and so, become at the same time actors. It is in n this intersection between public actors and private actors as directors and the social forces expressing themselves in an association of inhabitants with regards to the space, that one can make an interaction between the effective information of visible space.

He concludes the interview affirming that:

There is a new dynamic of social groups and there is the phenomenon of exclusion which passes over the position of the individual in the Marxist concept of class struggle. And there is the phenomenon of exclusion, which occurs in the domain of the research itself, when theorizing occurs. At this time, research studies the mechanisms of exclusion, of one group by the other, and precisely, these mechanisms are studied so that one can theorize.

Rochefort basically represents the phase of permanence and more intense renewal of the relations between France and Brazil in the field of geography. His first trip to Brazil, as a participant in the XVIII International geography Congress of UGI,[24] marked the beginning of a long period of activities in the area of urban planning and territory *amengagement*. Rochefort created a school of thought, with the National Geography Council of the IBGE becoming the center for spreading his ideas which were soon propagated across the country. What has been called the Rochefort method of urban analysis attracted a considerable number of renowned geographers of proven quality who had important roles in the departments responsible for planning, territory *amenagement,* and the management of space.

Michel Rochefort became a prominent figure in the academic world of geography. In France, at the time of the research, Rochefort remains a lecturer at the University of Paris-I, Sorbonne, where he supervises theses by students coming from far-flung parts of the world. He achieved a level of academic respectability due to his theses and practices of applied geography which granted him access to official and academic circles in France. He acted with such intensity in the planning departments, interacted with technical professionals of different fields of knowledge, gained notoriety and respectability, and at the same time became a great promoter of French geography. At Sorbonne, his professional practice was soon

---

[24]With regards to the participation of the French committee in this event see Valverde, O. "La cooperationfrancaise dans la geographiebresilienne", In: Cardoso, L.C. e Martiniere, G. op. cit.

recognized and valued. This fact can be proved in the enormous power of attraction he exercised over diverse areas of knowledge. The majority of the theses that he supervised came from professionals from other areas rather than geography. In Brazil, the older geographers knew his works well and had Rochefort as their theses supervisor.

His twenty-six journeys to Brazil demonstrate how close these relations were. An important fact to emphasize is the effect of Brazil, with its large space, with its singularity, on the formulations made by the eminent professor. The account of his experience in the northeast of Brazil, when "his concepts of rural and urban fell apart," reveal the value of this practice of exchange and sharing in the process of constructing scientific knowledge. Brazilian geography incorporated much of his reflections with similar situations. The opening up of horizons, the unveiling of forms, and relations are fundamental for the advance of science.

> Many geographers used the Rochefort method when they studied the aspects of the urban geography of different Brazilian areas. The results that they reached can express the functional reality of the space studied, to the extent that other elements were used to comprehend the life of relationships.[25]

Rochefort displayed a deep and detailed knowledge of Brazilian issues, indicating during the interview situations linked to the formation of the country and the condition and propositional level of Brazilian geography.

Rochefort's name remains as a great reference point in French geography yet these days one perceives a certain ostracism of Rochefort; he is more of a reference point. This does not mean that he still participates in academic events. However, he has been present in other cultural events such as his participation in programs with the dimension of *Les Entretiens de la villee* program, and in the *Espaces &Citoyennete* panel with the *Voies de communication et flux* conference which took place on April 7, 1993.

This situation is probably linked to Rochefort's extreme dedication to applied geography manifest in urban and territory *amenagement*. With regard to the presence or influence of French geographers in Brazil, there remains a certain replacement by other French specialists. The reality is that the country has invested very little in the last years in spatial policies with greater consequences, especially with regard to official projects by the federal government. This lacuna was filled by local technicians and successive cuts by the federal government have made foreign consultancies unviable. Even so, the importance of Rochefort cannot be neglected. To do so would be to damage the effective reading of the situation of cooperation which took place between both countries and which was strengthened from his first visits onward. His presence is behind much of the knowledge produced in the sector of urban geography. The select group that he organized and directed, the qualified members of the ministries, councils, and commissions in which he acted as an advisor form human resources of excellent quality which modified the profile of Brazilian production in the area of the urban and the regional.

---

[25]Correa, R. Lobato. op.cit. p. 190.

Bernard Kayser was another great influence upon Brazilian geography. He was part of the French committee which went to the UGI in Rio de Janeiro in 1956, and which would start a long period of exchange with Brazil. He taught various courses about Brazil in Toulouse, some together with Milton Santos. His expressive contribution appears in *GeografiaAtiva,* especially in the first chapter of the fourth volume, "The region as the object of study in geography." There he works with the concept of economic space and polarized space.

What follows is the outline of the answers received from the interview carried out with the eminent Professor Bernard Kayser:

I asked Kayser what he thinks about Brazilian geography if it has its own profile and constitutes an autonomous school.

My sympathy for your country and for Brazilians has no restrictions and it is only health reasons which have recently stopped me from continuing to offer my collaborations. It is evident that Brazilian geography exists. It has the same epistemological problem of geography in all countries - a "science" without a method, an objective; the work (often) of excellent specialists in their own field of work who institutionalize themselves as a reflex action to defend their profession. Brazilian geography does not exist, but the geographers do, especially Brazilian ones.

Discussing the presence/relation of French academics in Brazil, I asked Kayser if there had been discontinuity in our relations and what would be the symptoms of this stagnation. Kayser opined as follows concerning the different moments characterized as hegemony, distancing, rupture, and approximation in our scientific relations and the significant role of professors such as Deffontaines, Monbeig, Papy, De Martonne, Pierre George, Kayser, Rochefort, Labasse, Claval, Lacoste, Tricarte, and Pebayle, among others:

Your relations with French geographers are perfectly eclectic. You look in them for what is useful for you, and that is right. But contradictory influences are nevertheless another obstacle towards cohesion in the discipline.

In terms of outlining his intellectual itinerary by referring to the role that his work in Brazil had in his formation, he said:

If I had not worked in Brazil, I may have not developed the intuitive doctrine with regards to the Third World which led to some courses and articles, many times going against the current (disagreement) of common opinions and analyses.

As a recognized professional in academic circles, Kayser was asked about novelties in terms of geography in France. Speaking in terms of consistency, characteristics, the centers of production, and the main characters, he answered:

There are certainly many novelties in French geography. I can emphasize the commercial production of cartographic analysis and statistics, completely separated from the field of reality, the concrete and social realities. These are modern analyses with a publicity character, therefore profitable, income generating, as can be seen in the production of the "RECLUS" group. I would also single out the consolidation of the "social geography" current, which lives and is often militant in its contact with reality, modest, but efficacious, based much more on human effort than machines as can be seen in the texts of *Hérodote.*

4.1.6 Ruptures, Tremors: The "New Geography"—Yves Lacoste Does Geography, War ...

Yves Lacoste is, without a shadow of doubt, one of the most well-known French geographers in Brazil. His ideas spread through the pirated circulation of his book *A Geografia serve antes de mais nada para fazer a guerra*[26] and caused a great commotion in its Brazilian audience and influenced a whole generation. The distribution of Lacoste's work during this period and in the subsequent years went beyond the nation's borders, considering the expressive presence of geographers from neighboring countries participating in events promoted by the AGB.

In July of 1978, when the Brazilian Association of Geographers was promoting the Third National Gathering of Geographers, Lacoste's ideas became key slogans. This gathering was truly a turning point in the context of discussing, organizing, and mobilizing Brazil's key issues, as the nation still suffered the strong effects of the military dictatorship's policies. All the accumulated repression, the forced silence led to a cry of liberty which had its excesses, hurt many people, removing the leadership of the AGB, who were called mandarins, and resulting in the election of a new leadership. In a short time, a new group of mandarins took over the entity. All the discussion surrounded issues of re-democratization, openness, amnesty, and constitutions. The basic presupposition was that democracy should start within the entity. From this principle, a profound transformation in the personnel of the AGB was begun. Lacoste was in the background of this review of the AGB's practices and organization, as quotations attributed to him were on everyone's mouth leading to enthusiasm in some audiences and fear in others. One cannot affirm that all that was said, loud and proud, came from Lacoste, but what was said in his name, certainly.

The actions of the new leadership and their repercussions in the local sections of the AGB contributed to nationalize the movement and make it virtually one. New subjects emerged based initially on the words and key themes of the book. These included among others: the geography of teachers, the geography of greater states, region as obstacle, the problem of geography's epistemological deficit, the question of scale in reading, and the interpretation of phenomena. All this took place in the midst of a heated debate which did not end with the event in Fortaleza but led to an overhaul in the category, and gave teachers a greater visibility, a segment that did not have much expression within the AGB until that time. This led to a change in the association's rules, meetings, the proliferation of sections, and a greater vitality in the editorial market. New actors entered the scene, some of them not so new but who had distanced themselves from the country due to the repressive regime which had been implemented. This is the case of Milton Santos who will be treated separately in this book. But returning to Lacoste who, with his geography and his war, carried out a revolution in Brazil. In the scope of the contestation manifest

---

[26]Lacoste, Y. op.cil I.

during the event in Fortaleza, his proposal comes as a kind of antidote to all the ills which had come upon Brazilian geography up to 1978. This led to many exaggerated statements and affirmations, and absurd texts were written and spoken with little criteria with some using much verbal abuse and others omission in an extremely turbulent period. Lacoste did not expect his book to cause such a controversy, so much scandal on "that side of the Equator"... maybe, yes. The author affirms in the initial words of the preface of Maspero's second edition in 1982:

> As soon as this booklet appeared in 1976 there was a beautiful scandal in the corporation of university geographers, such a great scandal that many among them became full of indignation.[27]

Lacoste was already known by Brazilians. His books, *A Geografia do Sub-desenvolvimento,* published in France in 1965 by PUP, with a new edition in 1981 and *OsPaíses Sub-Desenvolvidos,* part of DIFEL's Saber Atual collection number 62, São Paulo, 1966, 3rd edition (published in France for the same collection under number 863 with the fourth edition in 1963) rapidly reached a Brazilian audience. In 1964, Yves Lacoste alongside Bernard Kayser and Raymond Giglielmo, under the leadership of Pierre George, wrote *GeografiaAtiva* edited in Portuguese in 1966 by DIFEL of São Paulo. This book caused a strong impact on the specialist Brazilian public.

The book was translated by four promising geographers from the USP geography department who achieved a great respectability in the Brazilian field of geographic production. The book's impact came at a time when postgraduate study was being established in Brazil with the development of more regular courses with a wider scope, and it caused a commotion in the spaces of scientific creation and teaching of geography especially through its concepts of active and activity. In terms of focus, in this part of the research Lacoste, as an important figure in French geography, blends in with the other authors of *GeografiaAtiva*. The action of the other authors will also be observed, especially Pierre George and Bernard Kayser. It is important to remember when considering the approach adopted and its impact subsequent to its release that the book was the work of a quartet of authors. This means that the references to P. George and his relationship with Active geography *GeografiaAtiva* we are referring to the work of these four authors. For the four French authors of this book, the objective of active geography is clarified in this passage by P. George:

> Impatient to affirm themselves as useful to regional or national social and economic development, geographers from diverse countries such as France, Belgium, Northern Countries and also from regions where the problems of space impose themselves more dramatically than in our old Europe, in Brazil, in the socialist economies such as the Soviet Union, Poland, Czechoslovakia, launched the idea, after the Second World War, of an applied geography, modeled after applied geology. It is, in spirit, an attempt to centralize the analysis of facts and the reporting of facts about themes which could contribute, in the

---

[27]Lacoste, Y. op.cit p. 1 I.

best time possible, to provide information for those services or those companies which have
the role of using or increasing the value of a fraction of the territory.[28]

Active geography became fashionable, attracted a large number of geographers,
and spread itself in scientific and planning circles. Its discourse met the needs of a
world which was transforming at an intense speed and which had a wide range of
distortions and needs that were unacceptable to more sensitive and politicized
persons. The period after the war created a sense of possibilities and an extreme
belief in science as transformative. Alongside the great technological advances that
were immediately communicated to the world (never at the speed witnessed today),
there was a philosophical position which guided political actions of a functionalist
structuralist mode based on the development versus subdevelopment discourse with
evolutionist, why not Darwinian interpretations. These were ideologically worked
actions, to the point of trying to prove that subdevelopment was a stage, a phase
which countries encountered on the path of development. Statistical averages were
regarded as having unparalleled utility. Indicators and variables appeared in the
discourse of specialists, among them those economists who were preparing to
launch the great ascension of their category based on different, sometimes divergent
schools yet oriented toward the same goal. Postwar liberalism fed all this illusion,
corrupted hearts, minds, countries, and nations. When the concept of a Third World
is established, one presupposes other worlds. The Third World is easy to perceive,
and it is the world of the poor, of political backwardness, misery, and subdevelopment. This is the period of the Cold War, the rivalry between the countries of the
typically capitalist bloc, characterized by the free market which places in conflict
the United States of America and the Union of Soviet Socialist Republics. The First
World is formed of the industrialized countries which provide a high standard of
living for their inhabitants. The Second World, those led by the USSR and other
countries of the so-called *IronCurtain*, is the name given to the countries which
experimented with so-called communism, marked by the excessive presence and
control of the State over the lives of individuals. It is Lacoste who offers an
excellent explanation for the expression "Third World," fundamental for the
understanding of the principles of active geography of which he was one of the
authors.

> The ideology of national movements in the colonized countries, the struggles that preceded
> their independence, the conference carried out in Bandung by the representatives of the
> African and Asian states and so many other facts reinforced the idea of a grouping of
> under-developed countries in a kind of alliance of demands towards the western countries
> considered to be the direct and indirect cause of their under-development. At the same time,
> it became necessary to develop a concept to designate this relative unity of under-developed
> countries, the human mass of which they are constituted, and their misery which is
> attributed to colonialism. This explains the birth and the success, at least in the French
> language, of the expression "Third World". It was developed by A. Sauvy, imitating the
> Third Estate of 1789, which in most of the nation was formed of diverse classes and social

---

[28]George P. et al. GEOGRAFIA ATIVA, São Paulo, DIfEL/EDUSP, 1966 (Translated by Gil
Toledo, Manuel Seabra, Nelson de la Corte and VicenzoBochicchio) p. 15.

groups who demanded the rights which until then were confined to the two other Estates, the nobility and the clergy.[29]

GeografiaAtiva, although linked to other presuppositions, has at its heart a super valorization of geographical discourse in stating that "It is not possible these days to carry out a good administration, in a public or private scale without a solid geographical culture or a selection exam in geography," while at the same time its content is very critical toward geopolitics affirming that "the worst of the caricatures of applied geography in the first half of the twentieth century was geopolitics, being used to automatically justify any territorial demand, any taking of land using pseudo-scientific arguments." This book, which was full of innovations and was responsible for a change of epistemological attitude in Brazilian geography, provided an interesting discussion about the object and the methods of geography (item two of the second part of the book). The authors are emphatic, based on six points:

1. Geography is a human science.
2. Geography is a science of space, but its methods are different from the natural sciences of space.
3. Geography is the result of and the extension of history.
4. As a historian of current times, a geographer should continue the studies of a historian, applying their own methods.
5. The objective of applying geographical methods is the knowledge of situations.
6. The study of a situation may proceed from a contemplative conception or an active conception.

Another focus point of the book which became a reference point in Brazil was the discussion by the authors on the dependency of French geographical research, seen by then as organically associated with the university teaching of that country.

The purpose of this discussion about the book *GeografiaAtiva* is to show the origins, practices, and affiliations of the so-called French school of geography. For this purpose, it is important to emphasize the analysis made by P. George, author of the first part of the book, but which is present in the whole book: problems, doctrine, and method with regard to the "organization of geographical research in France; its role in the development in active geography" present on p. 41. He immediately points out:

> Geographical research is until the present organically associated to university teaching in France, there is no center or geographical research service that is free from teaching responsibilities which whilst well attended consume significant resources. This means that they are not able to diagnose situations in a continuous manner. It seems certainly to be desirable that a research organism be formed with a dual vocation of fundamental research and geographical formation, being called, according to necessity, to collaborate in regional or urban planning enterprises or market studies comparable to the Centre of Sociological

---

[29]Lacoste Y; Geografia do Subdesenvolvimento, São Paulo, DIFEL São Paulo, 1975 fourth edition, pp. 17–18.

Studies of the National Centre for Scientific Research. All the works carried out by university geographical institutes provide a contribution of great scientific value and much practical interest while the expectation remains that centers of regional study and a center of national scope be in conditions to meet the whole set of needs that exists. These works refer to two levels of scientific formation and inquiry. The first is researched carried out by beginners organized and controlled by a research director. The second level is occupied by research carried out seeking to prepare a doctoral thesis. The lack of knowledge of the interest of these works at a practical level comes from two causes: the first is the specific difference between the geography taught through school programs and geographical research at university level and that of fundamental research... the second cause of the misunderstanding comes from the disordered statements of those who preach a piecemeal geography and suggest implicitly, or sometimes explicitly that geographical knowledge in general is "not applicable and with no interest for the management of private and public goods.[30]

The whole text is characterized by the valuing of works of geographical research while at the same time revealing the author's utopia in search of the recognition of geography as scientific knowledge with a recognized application. Planning logic permeates the book so that on p. 23 there is the affirmation that "the objective of the application of geographical methods is the knowledge of situations." In the text, it is evident that the knowledge of situations corresponds to the diagnostic evaluation, the first phase of planning, preceded by basic legislation and propositions according to how it was conceived and executed at that time. It consisted methodologically and ideologically of the belief in planning as the solution to the crucial problems of development, dividing the analysis in three moments beyond the diagnosis. The second phase was that of basic legislation and propositions.

One cannot deny the fact that the authors were militant in and experienced a great influence of the French Communist Party (PCF) despite the rupture that took place in 1956. The concepts of a planned economy allied to the knowledge that P. George had of the reality of the former USSR reinforced the position expressed in the book. The great thrust of the book and its proposal resided in the questions linked to development, which were Lacoste's responsibility (pp. 47–158) and developed in his introduction to the second part of the book, with the title "Perspectives of Active geography in Underdeveloped Countries." This was divided into two items—The greater geographic situation: the Third World and a geography of discordances. At the end of this second item, the author is emphatic and makes it clear what will be the genesis of his small, very successful book *A Geografia serve,*

The timidity with which geographers have participated until now in the study of underdevelopment has complex causes. This may be an effect of the successful study of the geographical realities in developed countries which are often in equilibrium. The geography of the Third World is, to a large extent, a geography of discordances and disharmony. It is also important to mention that not all the obstacles that have stopped a greater participation of geography in the study of underdevelopment are related to geographers. The imperialism of certain economists and their lack of knowledge of geography have not made things easy.

---

[30]George, P. et alii. op. cit. p. 41.

Furthermore, the lack of precision, the relative character, if not objective, with which numerous theoreticians approach the concept of underdevelopment do not favor its usage by geographers, who are linked to the study of concrete realities.[31]

Lacoste makes it clear that there is a need to study underdevelopment in the context of another perspective, one which permits the insight of a geographer, abolishing the corporative domination of other specialists. It registers simultaneously the difficulty of the geographer in adopting concepts of underdevelopment that are distant from explaining concrete realities. It becomes evident that at the same time in which Lacoste seeks to defend geographers, he makes clear the theoretical fragility of the category which has empirical reality as the basis of its formulations.

While the professional practices of geographers such as fieldwork, accounts of scouting expeditions, and/or exploration are characterized by the strength of the empirical approach, various extracts such as the one below indicate that they are not the major characteristic of the text which stresses critical rigor, even though Lacoste himself reveals empirical strengths:

The social relationships which existed in most of the Third World and especially in the rural zones are of a particular nature, very frequently ignored by political economy treatises. In countries which are now underdeveloped capitalism was abruptly introduced from the outside by the actions and for the benefit of a colonizing or a native minority in a dominated society in which other economic and social relationships prevailed. This minority, taking advantage of the political debility and the technical backwardness of the submissive populations, became able to operate a true perversion of the normal mechanisms of the capitalist system; the already considerable powers of those who owned capital were reinforced and transformed in monopolies without restraint by a historically monstrous linkage with the form of dominions exercised by feudal lords [...] The independence of the majority of colonized countries has not led to the disappearance of the monopolizing minority, as native minorities either replaced or joined the foreigners.[32]

The presence of Lacoste in active geography is significant. The author constructs his argument based on the themes of Third World and underdevelopment, resources for geopolitics, a favorite subject to which he dedicated himself later and which these days form the basis of his geographical formulations.

In 1962, Lacoste wrote "Underdevelopment, some significant works appearing in the last ten years," in *Annals de Geographie* March–April, July–August volume. In this article, the author clearly indicates his option for the theme that would offer him various opportunities in the editorial market and would provide the conditions to develop a more consistent and consequential criticism for geography. In the epilogue of the French edition of 1982 of *A Geografia serve...*, Lacoste revises his criticism of the work of Vidal de la Blache. The epilogue reveals the author's significant maturity in a period in which he shows himself to be less inflammatory

---

[31]Lacoste, Yves. op. cit. p. 51.
[32]Lacoste, Yves. op. cit. pp. 64–65.

and when he is already in dialog with professionals in other categories, with the military, etc., through the Hérodote journal.

Currently, Lacoste is a professor at the University of Paris-Saint Denis, Paris-8 where he is the director of the Research and Geopolitical Analysis Centre. His thesis, examined in 1979 at the University of Paris-I, was on the theme "Unity and Diversity in the Third World" having M. Rochefort and P. George as part of the examining panel.

In 1993, his research was on the following subjects:

- Geographical epistemology, geopolitics (internal and external);
- Problems in the Third World and Post-Communist societies,

He is also the director and one of the main supporters of the Hérodote journal. It is useful to emphasize some passages from the editorial to the first edition of the journal released in January 1976, which have a significant contribution from Lacoste. The title of the editorial on its own attracts interest—"Attention - geography!" After the introduction, five themes are dealt with—Attention: geography informs the major states. Attention: geography becomes mystical. Hérodote: the inaugural contradiction. Learning to think about space to think about power. From the critique of maps to the critical letters. Throughout the text fundamental issues for the development of geography as scientific knowledge are dealt with. It begins by saying that words and images proliferate in geography and that they contaminate language. It affirms that everyone knows that space is finite, and that it can in fact be expensive, and polluted. He is strenuous when he says "*social relations are inscribed, they impress themselves on the scenery as on a registering surface: memory.*" In the final pages, after questioning geographers and challenging them regarding their images and words he concludes emphatically: "*To criticize is to place oneself in crises. To be polemical is to make war.*"

With regard to his writings, beyond the books already mentioned Yves Lacoste has managed to publish and disseminate his research widely. In 1980, he published his thesis, *Unity and Diversity in the Third World*, through Decouverte in Paris with 526 pages. In 1986, it was the turn of the large work surrounding *Geopolitics of the French Regions* edited in three volumes with a total of 3500 p. by Fayard, of Paris. This work led to prominence and respectability for Lacoste, who coordinated a reputable group of professionals to carry out the research. In 1988, Le Livre de Poche de Paris published *Question de Geopolitique* with 250 p. and in 1990 *PaysagesPolitiques* with 285 p.

All this production sustains the reputation of the geographer Yves Lacoste and gives him the status of an intellectual engaged in the major issues.

In Brazil, there is no logical order to his most well-known works, their editions being dispersed. While *Ospaíses sub-desenvolvidos* reached a wider, less select public, still thirsty for information about the theme, the book *A Geografia serve antes de mais nada para fazer a Guerra* also reached a wide public, mainly identified through their political links with the Left. The number of left-wing

Brazilian geographers, many militant, who followed Lacoste's ideas and used them in their works, is high.[33]

Lacoste's presence in Brazil became stronger after the aforementioned AGB meeting in Fortaleza. His importance and the influence of his ideas in Brazil is such that his status as a reference point remains, whether or not his books have actually been read, especially in the case of *A Geografia serve....* Lacoste was interviewed on January 7, 1993. His points of view and analyses were of capital importance for the elaboration of the research which formed the body of the book. The interview, initially arranged by telephone, took place at the University of Paris-8, Saint Denis. Lacoste received us in a sympathetic and attentive manner creating a pleasant environment which facilitated the flow of the interview. This environment allowed us a better understanding of the illustrious professor and also provided elements for a critical assessment of the French influence on the formulations, practices, and questions of Brazilian geography.

4.1.6.1 Interview with Professor Yves Lacoste

The conversation started dealing with Brazilian geography after 1956 when the International GeographyCongress took place in Brazil with a strong French presence. From this date, French influence on Brazilian geography was reinvigorated. Lacoste had little idea of his importance in Brazil, especially through his book *A Geografia serve.* When asked about the importance of his book in the formation of Brazilian geographers, he answered:

> No, I was not aware, but when I became aware of this importance I felt much honored. It is actually necessary for me to completely re-write this book. It does not mean that I am against the ideas I launched in this book, but I have progressed a lot. This book is from 1976, and was born in Paris with the birth of the Hérodote journal. This journal still exists and has an important role in the European context being printed industrially. Initially, Hérodote's concern was that which I expressed in my book, dealing with development and a deeper look at the subject. After, from 1985, there was a change for until then the journal had been critical of the academic discourse of geographers, taking into consideration all the ideological, strategic problems, remaining faithful to its title which has a geographical and ideological strategy. It remains faithful to its title, its sub-title, but after 1985 changes began in Eastern Europe with Perestroika, etc. Becoming aware that this change would be considerable we became interested and dedicated an edition of Hérodote to the Soviet Union for the first time, I believe, in 1986. Until then nothing had been covered in Hérodote about

---

[33]For a list of the various Brazilian geographers who cite and refer to Lacoste in their research, see Milton Santos NovosRumos da Geografia Brasileira", HUCITEC, 1981, Oliveira, AriovaldoUmbelino de "Espaço Tempo - Compreensãomaterialistadialética", Moraes, Antonio Carlos Robert e Costa, Wanderley Messias "Geografia e o Processo de valorização do espaço, Gonçalves, Carlos W.P. "Estrutura Agrária e Dominação do campo", Andrade, Manoel C. de "0 Pensamento geográfico e Realidade Brasileira", Santos, Milton" Alguns problemas atuais da Contribuição Marxista àGeografia, Santos, Milton "Por uma Geografia Nova, São Paulo, HUCITEC - EDUSP, 1978, - Moraes, Antonio Carlos Robert. GEOGRAFIA - Pequena História Crítica, São Paulo, HUCITEC, 1981.

the Soviet Union because it seemed to us to be absolutely static with nothing new. From 1985/86 onwards there were some changes and since then geopolitical problems have emerged which are more and more difficult and at the same time more and more contemporary. For example the following edition of Hérodote, which required a lot of effort, was dedicated to the problems in Yugoslavia. On the whole, between 1967 and 1985 we were mainly critical, saying that geographers and others- economists, sociologists- did not deal with the problem based on an effective analysis of geographical information. We did this for some time. Happily, after 1985 things started to change, a critique is made of how older phenomena were explained and now we try to explain problems which develop quickly and require much more understanding. We are trying to explain battles, and we do not know how they will develop, who will win them, who will lose them and now we advance, we are in a more direct observation of the problem.

Continuing with information which could provide context for the interview, the issue of the language of *A Geografia serve* was focused upon. We affirmed that the book had a language which is very close to that used by Hérodote, a mixture of discourse which reveals his direction of the journal. This book was published in Portuguese, in Portugal by the IniciativasEditorias de Lisboa publisher in 1977, and was only officially published in Brazil three years ago. However, a group of geographers who led the AGB (Brazilian Association of Geographers) from 1978 onward made a pirate edition of the book. There was no other solution. Every time we sought to edit the book according to the legal norms, we had problems with the Portuguese publisher. In the end, the pirate edition was a solution and not a problem. The book was consulted and multiplied, photocopies were made all over the country, especially the chapter "The geography of teachers" which provoked a great discussion in the process of transforming the teaching of Brazilian geography. About this theme, Lacoste stated that "I was always the object of pirate editions. *The geography of Underdevelopment* was translated into thirty-five languages and there were twenty-five pirate translations."

I informed him that we were forced to do this (laughs). I also stated that there is a gap between the discourse of *A Geografia serve* and Hérodote, which is very specialized, and is aimed at a restricted public. Lacoste agreed with me. Continuing I asked if there was indeed a gap between the times the book was written in, the discourse included in it and these days. I affirm that many Brazilian geographers expect something from Yves Lacoste, that there is a gap something to be filled. Feeling provoked, Lacoste responds in this way:

> Your observation is very subtle, very just. When I wrote *A Geografia serve* it was a critical reflection on teaching, on the geography of teachers, but gradually, the Hérodote group progressed, it was not just me; we were a group. I started with young women who were postgraduate students in 1976 and today they are friends and effective specialists and at the same time we progressed. While it is true that Hérodote had become a very focused journal, as you said, for a specialist audience and mainly an audience of journalists who were concerned with geopolitics, politicians, military leaders, businessmen, etc....

I continued with my provocation and said "and also of geographers, sometimes." Lacoste soon retorted:

> No, because university geographers do not like this... and I think that you are right. It is necessary that I write a new edition of this book, *A Geografia serve* to do a bit of a

theoretical and practical outline of the progress we made. At the moment, with the Hérodote team I am compiling a thick edition, the "Dictionary of Geopolitics", which will come out this year, September 1993. It is hard work. After this I plan another book to come out in 1994. I want to tell you that this book will be written from the other side of France ...

I try to guess and ask "in Algeria." Lacoste answers, "No in the Antilles." I retort "is it also a research focusing on the theme of geopolitics." Lacoste answers,

Yes, it is research in geopolitics but because I am going to Martinique to do this research and afterwards the political situation became very difficult because in the place I am going to do the research a local politician died suddenly and therefore I can verify the structures of patronage, of planning better and alongside the observations I am going to make, I will do something, I will finish writing the book.

I interrupted saying that this is very good, but is a bit American in style. Lacoste retorted "It is to a certain extent American. The title, *A Geografia*is a very adequate, faithful title and is a reflection, I remember very well, accompanied by some glances at Europe." In response to this, I asked "Can it be said that the geopolitics with which you always worked is a new regional geography which is being defined more precisely today?" Lacoste answered as follows:

Your question is truly a very interesting question; it is a very subtle, very deep question. I will answer yes and no. Yes, I think it is possible and necessary to establish the foundations of a new regional geography on geopolitical analysis and the political influence of this or that person, of this or that policy, there is a geographical aspect and an important form of organization of space. I am not the person who is the most advanced in this domain; it is Beatrice Giblin, someone who carries out an important role. After some time at Hérodote she has written a book which is at the same time theoretical and practical and which is called *The Region, political territories* I emphasize that it is political territories in the plural. She presents the Northern region of France as an example that she knows very well. And what is interesting is that she did this book for her thesis, and she did it by interviewing many politicians. It is the first time that a geographer who comes from this region has met with politicians who have told her many things. This book sparked the interest of politicians, and also of students on a French law course. It also attracted the attention of a political councilor. She has had excellent contacts with politicians even those from different and rival parties, including those with a rightwing geopolitical persuasion. She is currently responsible for doctoral formation in geopolitics which involves about fifty students who come from different horizons including Brazilians.

Motivated by my interest, the professor continued:

There are many students who come from large schools and there are many geographers. She is responsible for the seminars which we call, for simplicity sake, *Internal Geopolitical Vision* because it deals with the geopolitical problem inside the State and she supervises a certain number of theses which are very interesting. Together, on the other hand, we published a book; you might know it and it is called, *Geopolitics of French Regions*, where we gather about forty geographers, forcing them to work a bit...

I tell him that I know the book and briefly mention its content which deals with electoral geography. Lacoste continues the conversation saying:

Yes, it is the book Electoral geography with 3100 pages. It is a regional geography approach. So I answered your question with a yes. But now I answer *no* to your question because in the same geopolitical game there are issues which are not part of regional geography but which are problems of movement, of action, of military operation. We are extremely interested in military problems, which take place between two forces that are opposed in the field and which are concerned with problems in the organization of space. You have the permanent forms of organizing space, civil society, economic, political and social activities. Well, this is one level; the other is that of war, because the things of war do not interfere directly in the regional geographic plan, if there are small military operations in the regions. This is the *no. Yes*, geopolitics is the basis of a regional geography, a geopolitics or geostrategy, but not of a strategic geopolitics which is something else.

I ask if this is a new field or from his exposition in this moment when the end of History is discussed, could one also talk about the end of geography. We laugh and Lacoste complements "Or the opposite, the triumph of geography." I retort and ask whether this would be the triumph or the end of Vidalian geography. Lacoste responded firmly saying "No, no, it is not Vidal's geography. I do not know what issue of *A Geografia serve* ... you saw!"

I state that I know both, that I have indeed read the edition with the epilogue. Lacoste retorts

With the epilogue. At the same time it may be the end of Vidal's geography for that part of French geography which used Braudel to incorporate university geography. It can also be said that what I do is also the triumph of Vidal's geography because there is a book that is completely impregnated with Vidal de La Blanche, *The France of the East,* which did the opposite of geopolitics.

I say "professor, you wrote the text 'A bas Vidal ... viva Vidal!' about the book, I read in Hérodote, and you say that *The France of the East* is a forgotten book and I ask for this reason it is traditional?" From then onward, the conversation takes on a more intense dialog. Yves Lacoste responds "I think that we have the real triumph of geography." I ask, whether this would not be linked to the so-called crisis of modernity, a crisis of history? Lacoste soon retorts.

No, no, that has all been overcome. Geography, in my opinion is now coming out of the blockage placed upon it, it progresses well, evidently with terrible things, conflicts with the group which I consider the most scientifically advanced in geography, the RECLUS group. We do not have the war officially, but I think the declaration of war will be made in the following month (laughs).

I provoke the professor questioning why it is always said that what is new in France comes from the MGM (Maison de Geographie de Montpellier) and I ask, what is really new in France in terms of geography? Lacoste expresses himself as follows:

What is new can be negative (laughs). What seems to be new, what is seen as new, is very bad. I can explain a bit. It is triumphant like all other geographies. We currently have great problems such as the geography of the East. As you should know, Soviet geographers faced a terrible situation, because after 1941 human geography was forbidden in the Soviet Union. After 1941 there was no longer any human or economic geography in the Soviet Union. It was a consequence of the geopolitical problem, as it was a consequence of the German-Soviet pact that Stalin and Hitler agreed upon in 1939. Stalin believed in this long lasting alliance with Germany because he believed that the Germans represented this attempt, as a great geopolitical theory, the only one which would make a great continental unity. This was Mackinder's famous thesis, which was re-adopted due to the idea of the geopolitical game undertaken by Hitler. Stalin believed it. Stalin did not predict that two years later the alliance would lead to an attack. And he was so serious that, as Mackinder was a geographer, a specialist in human geography, Stalin believed that human geography was a diabolical, imperialist invention and from that time forwards human geography was forbidden. And this had repercussion in the other socialist and pre-socialist states with the exception of Poland. Polish geographers had close contact with French geographers to keep human geography. For example, I had a close contact in Cuba and Vietnam, which were extreme cases, where human geography is forbidden and people that I know will go to Cuba as missionaries and human geographers, I ask of human geography, they tell me they have to say that they are sociologists, historians. And in Vietnam human geography is forbidden. Vietnam presents us with a specific case because it is not the general secretary of the party who is in charge but the head of state, and the head of state is Ho ChiMin's successor. In 1973 I intervened with a research on the bombing of ditches which allowed the role of the geographer to be seen for what it was. In this I had a very curious role, which was when I spoke to my companions - in the socialist states they never left us completely alone and I had twelve companions - they were physical geographers and I was interested in human geography. So, consequently, we did what human geography does, and, unofficially, I was kind of a monitor of Vietnamese geo-morphology, so as to be able to do human geography (laughs). Currently we have, for French geography, important contacts with the Soviet Union[34] – with Russia, and there will possibly be a Hérodote journal in Russian and on our doctoral courses we have a certain number of young Russian geographers who come from over there and we prepare them so that they become good geographers. We have, therefore, a lot of work!

I asked Professor Lacoste, "with regard to the supervision and preparation of theses, I do not know if there are others, but according to the research that I did, you have only supervised one thesis by a Brazilian, Resende Dantas, in 1972 at the University of Paris-Vincennes, the title of the research was "Forms of Urbanization in Underdeveloped Countries" and I could not locate any other Brazilians. This is an aspect I want to discuss, as I do not understand, how is it that you have such an important role in the formation of that generation of geographers I mentioned, and when we talk about the supervision of theses it is other professors who are traditionally the supervisors of Brazilians. What do you think about this?" Yves Lacoste answered as follows:

Well, I can explain. There are two reasons. The first is that I did my doctoral thesis late, in 1979. Very late. Although I was a well-known geographer for quite a while, I did many other things and the fact that my thesis had its own particularities meant that I have only had four experiences in supervision. I was not officially a professor. I carried out the function of professor, but was not a professor. I could not officially supervise a thesis and for this reason there were a certain number of people, many Brazilians- who asked me to supervise them and I would listen to them but tell them that first I had to supervise my own. The second reason is that in the institution of the French university there are certain positions which in the past were considered scandalous, etc. and the people who worked with me in the past faced prejudicial consequences and ... everyone "disappeared" and I also "disappeared." The support I received was from the Hérodote journal, which never stopped giving me official credit and allowed me to request this credit from the geography

---

[34]The Union of Soviet Socialist Republics ended in 1991. In December of that year, the Community of Independent States was formed with various countries with economic and political connections.

Commission, etc., etc., and the journal never ceased giving me recognition. And now there is also financial recognition and this means that in financial terms the journal is well, with its editor Francois Mespero. My name was powerful and highly toxic. These are the two reasons. In this I answered your question. I have a good friend, Michel Rochefort, with whom the relationship between Brazilian and French geography is very close, and without doubt, there are Brazilian geographers whose names I do not remember who asked me to supervise their theses and I forwarded them to Michel Rochefort. That is the explanation. There is no doubt that I am interested in Brazil because I worked a lot on Cuba and from Cuba the specific problems of Latin America interest me very much. And if I have yet to go to Brazil it is not because of indifference towards Brazilians but due to different circum- stances. I worked on North Africa and then it was Tropical Africa and following that Southeast Asia where I specifically studied Vietnam.

I then asked, "Professor, about the Mediterranean, do you have anything to say? I read some research, a book of yours on Braudel, entitled *Braudel, a geographer*." Lacoste does not answer, he just confirms, "Yes, yes, Braudel, geographer."

Continuing the interview, I made the following statement "Here in France we can notice a type of shame regarding the past, people say 'I was a Marxist' or 'I am not a Marxist.' They convey sadness, some regret when they say 'I was a Marxist.' How do you see this issue, considering the importance of Marxism for critical geography? What do you think of this relationship—the fall of Marxism and the future of critical geography?"

Thus, instigated Lacoste manifests himself "You ask excellent questions. You have very good questions. I was a member of the communist party from 1948, when I was nineteen. I was very young. I remained until 1956." I asked if this was a consequence of the invasion of Hungary (laughs).

Lacoste affirms immediately,

No, it was not Hungary. Not Hungary. When I entered the communist party my political formation was null, null. In 1946, I can well remember, I did not even know what left and right referred to (laughs). It was null, completely null. Why? Why? Because, I realized, it was all explained in my colonial infancy. I am a colonial. I spent all of my infancy in Morocco. My father directed petroleum research. I am a product of imperialism (laughs). Initially I thought it was fun being a colonialist, a little colonialist, because I think, that up to a certain point, the role of colonization was not just negative, it was also very positive. And when you speak about Latin Americans, if you want to know, when I discuss with my friends from Cuba, who are all products of colonization, we feel very well together. After a certain point the colonial period was over, it became bankrupt and many French people in Morocco and other parts understood this well. Things happened differently in Algeria. So, in Morocco, when I was a child for my family this idea of left and right did not exist, or at least we did not talk about it from the start of the Second World War, young people like me did not talk about left or right, we were all against the Germans. In fact, soon after the war my father died and he was no longer around to explain things to me. So I joined the party with the support mainly of M. Rochefort and B. Kayser and all those who were younger than me but who, in my eyes, had considerable prestige, I considered them as greater than myself, I saw myself as small, I was small, very small. I remember the communist party; everyone today says that it was hard, controlling, doctrinal, etc. Three months after I joined the party I was made secretary. I didn't know anything but other friends told me that the position of secretary was always carried out by the latest arrival. It was there that I did my learning of a certain number of things and this gave me my construction of the world, my rationalization of the world. This added many things and I did not stop. And also, all my masters, my friends like Jean Dresch, P. George etc., etc., were all members of the communist party, so consequently I have a family history! (Laughs). And when I spoke using more elementary reasoning, I was not seen as responsible in the political sense of a militant, but as a friend, always as a friend, not in the political sense but in the geographical sense, very good friends.

Well, as I had lived all my infancy in Morocco, I wanted to go back to Morocco to do my thesis and they told me "Morocco is very disturbed, that the independence struggle had started, so I could not do my field work there. And Algeria was calmer." And I went to Algeria, at the beginning of the war and soon I became a member of the Algerian communist party. I did not know anything about Algeria but as I had come from France and as a member of the French communist party I became a member of the Algerian communist party and I was responsible for the party's intellectuals. One of the characteristics of the communist party is that it was maintained by the French in the country as the Algerians had been excluded two years before because of a small-bourgeoisie nationalist deviation (laughs). Yes, they were all French; there were military personnel, etc. It was interesting! And one characteristic of that group was that I was not linked to them because I was a geographer or historian, but the close link was political, militant and as an older professor I organized a group about colonization, about Algeria, about Morocco, about anti-colonialism and I did it with tranquility because they all knew that I was a member of the communist party. I did not hide it. I must point out that when the war began to become serious I was immediately expelled. At this time I had the opportunity of being expelled, otherwise I would have been in a dangerous situation and in 1955 I found myself nominated as an assistant at Sorbonne, with my companions. Do you know why I left the French communist party? It was not because of Budapest. They told me it was a revolution, a counter-revolution, I understood because at that time I already had a well-developed political perspicuity, although more elementary today (laughs) I was very young but the main reason was that the PCF voted in favor of a special power to carry out the war in Algeria and it was a leftwing party. In 1956 it was in power and so as to not lose Algeria the socialist communists and socialists came together to vote. It is easy to make accusations. They made the correct decision to vote in favor of the war and the communists also voted. And at this time, the discussion about the revolution inside the PCF became surreal and I politely withdrew. But in the end it became fun because in those days the communist party was considered renegade. All my friends who belonged to the PC and especially professor Jean Dresch, who was a great party intellectual, he always told me to remain calm, I would not be considered a renegade, and it is true that many years after leaving the party I was invited to become a sympathizer (laughs). When I made my preliminary annotations, because I was very young, I already applied foundational Marxists principles and slowly I started to realize that there were things which were more complicated than the official Marxist discourse as it was presented. My great intellectual torment was when I began to discuss, in 1962 and 1963, under the influence of Althusser, the Asian means of production because in my intellectual itinerary nothing had been more important than going to Algeria, going to the Maghreb which revealed great historians to me and others and until then I had followed the elementary Marxist path, the means of production, primitive communities, slave communities, feudalism, etc. And soon, thanks to Althusser, of course Althusser did not write about the Asian means of production, but it was he who one day in 1962 helped to organize a meeting (I invited him because I had left the party) and he brought texts by Marx, which had not yet been translated, about the Asian mode of production, I mean, a completely different periodization from what was known, and in a new perspective, much freer and much more complex, but, still, Marxist.

I would not call myself a Marxist, but I never absolutely denied what Marxist analysis brought me. I obtained many things which I would consider neither secondary nor essential. One of the things that concerns me now, and which already concerned me before, is the concept of the nation. For Marxists the nation is a difficult problem which is precisely in

contradiction with the distinct idea of a conflict of classes. For me, the nation is a fundamental phenomenon and in Geopolitics, in the present moment, it is even more important to
reflect on the nation when one sees what is happening in the Soviet Union, Central Europe,
etc. But as I was an anti-colonialist from the beginning, I am interested in the problem of
the nation. Therefore I have a position which is much closer to that of my students. Today I
am much more aligned with the French, I am a French post-nationalist and the nation, for
me, is something which is increasingly more important, that I consider necessary, more than
any other. It is the interest I took to the Algerians of being another nation, it is very
legitimate, it is very just and so, for me, the internationalism the Marxists speak so much
about is not the negation of the nation. With certainty, it is possible to do extraordinary,
magnificent things and for me, internationalism is the respect and the recognition of the
nation. I criticize my nation if it does things I consider evil and I address people who are not
from my nation to help them to be part of their nation.

I ask the professor if the sense, the concept of nation and nation state, would be
as he wrote for the journal *Parallele de Nanterre*, where he analyzed questions
linked to the macro- and microlevels concerning the *Europe of the 12* and the
struggle in Yugoslavia. Lacoste answered as follows:

Yes, I believe that the nation, for me, is certainly the domain of representation. In the new
research that we have it is very important because it is the idea that is being made and
representation is a very powerful, very strong concept. Beyond what is imagined, the
intellectual and moral values are linked to this and I believe that it is a way of conceiving
the nation, it is how the nation emerges, develops, spreads and sometimes disappears, etc.
This is my position with respect to Marxism, against the idea that it had of an ideal society
in which everything was determined for the evolution of the means of production. I have
not believed in this for almost thirty years, and consequently this idea of an ideal society is
an ideological representation.

I ask if this would be a form of utopia. Professor Lacoste agrees completely,

It would be a kind of utopia, but one which makes it necessary to analyze this society in
terms of the ownership of the means of production. I could say, yes, but it is not only this.
I believe that the Marxist analysis is necessary, indispensable but it is not sufficient; it is
important to take into account other things, the problem of power; the group which has the
power is one of the elements of the issue, even if it is not the owner of the means of
production.

I carry on asking if this would be power in Foucault's sense. Lacoste answers:

Yes, yes. The problem of power as a consequence of struggles. And one cannot compre-
hend a society, a communist state, if one does not take into account the terrible struggle for
power. Power is like a game, be it to control a bank, or control petroleum concessions, it
will always be necessary to struggle for power.

I continue asking if it would be power in itself and Lacoste quickly responds: "In
itself, cultural power also leads to struggles, there are groups." I ask if this would be
the representation of power and Lacoste soon replies: "Yes, it is the representation
of power. I can define the Marxist analysis as always necessary but not sufficient."
I take the opportunity to give the interview another direction, trying to obtain the
maximum information from the professor. I ask if he has many friends among

Brazilian geographers, if for him a Brazilian geographical school exists. Revealing little intimacy with Brazilian geography, Lacoste answered thus:

> There may even be one, but I do not know it. The Brazilian geographer I know the best is Milton Santos, but Milton Santos, I told him that he was much more concerned with understanding and quoting Anglo-Saxon and French geographers than talking about Brazil. It was logical because he was not in Brazil. Consequently, in another moment in an interview with Hérodote, he affirmed that he was an émigré geographer, he was not in his nation and for that reason spoke about other things. Everyone held Milton Santos in high regard, but after he returned to Brazil I did not see him again.

In the final moments of the interview, I added that I was considering for purpose of analyses, the role of French book authors and theses supervisors as mediators of French geographical thought along with those French scholars who researched or research in Brazil. I took the opportunity to single out Lacoste's case saying that it was very interesting as he was an author who did not know Brazil, but had played an important role in the formation of our geographers. Lacoste answered:

> It's funny. I do not know this country and how did I influence this country? (Laughs) It is necessary to know Brazil. I think that is the next step. Brazil seems very interesting to me due to my geographer friends, but Cuba also interests me because the problems are very large and very interesting.

The interview immediately reveals another Yves Lacoste, an author who does not seem to want to revisit his work, specifically the reference book, as if it caused harm. One cannot neglect the fact that the book *A Geografia serve...* is from 1976 and as is the case of any geographic researcher or theoretician; its content is dated. What was written was relevant for the days when it was written. Society changes, authors change, hence Yves Lacoste changed. It was common to characterize him through constructing a profile based on what was available in Brazil. Here, Lacoste was known through those works already mentioned, where the theme of underdevelopment was the major theme. In writing *A Geografia serve...*, Lacoste achieved a notoriety which is seldom achieved by a geographer in France. The book marked a change in the author's main concerns.

Until 1976, his work was produced in a mainly individualized form. Even in *GeografiaAtiva* which was written and proposed by the four geographers already mentioned, Lacoste focused on his preferred theme, underdevelopment.

However, at the same time that the book was released in 1976, the Hérodote journal appeared, the result of a collective work involving a considerable number of researchers. The journal gave visibility to the author who knew how to administer the historical context in which it appeared and at the same time introduced something new into French geography, a kind of reaction to the established geographical knowledge. The book as an individual work and a journal expressing the ideas of a group and the subsequent production of his doctoral thesis created the scenario in which Yves Lacoste has been an able actor. The creation of the doctoral program, the widespread acceptance of the magazine worldwide, and a strong intellectual investment in the magazine's subthemes (strategies, geographies, and ideologies) gave the professor an enviable position in French academic and editorial circles.

Lacoste intelligently followed the general movement of society, the world, and at a certain moment adjusted his focus to the salient questions of that historical moment. His acquaintances and alliances and his experience made him able to know the right path for the discussions and the right moment for a change of focus.

And where is Brazil? Nowhere. The interview proves that Lacoste, the specialist in geopolitics, does not identify himself with the Lacoste who constructed a formidable fame in Brazil. He has a notoriety which has been greater, faced with the changes in the world Brazil would not be excluded.

Lacoste still has his place in Brazil. His book, which became famous here, provoked a greater discussion among those who did not read it than among those who actually knew it. It was truly a war cry. On many occasions in the interview, Lacoste avoids talking about the book referring to the need to rewrite it, a form of denying his own work. There was no justification in terms of the context in which it was written; it was, in fact avoided, ignored. The interview allows the recovery of the illustrious professor's trajectory, his infancy, youth, the Party, the admiration for colleagues who were fellow party members, for the party and especially for geography. It also reveals his rupture with the party, his militancy in Algeria, the construction of his experience in underdevelopment, in the fight for national liberation, colonial action, etc. Lacoste discovers France and Europe, two important themes to acquire the visibility to declare new wars. In the text of the interview, he promises war, the enemy in sight: The RECLUS Group. Let us wait and see!

Lacoste's success in Brazil as a geopolitical geographer is far from that achieved through just one book, almost a catechism for a group of fanatics who imposed its reading and transformed part of its content into slogans. His book *A Geografia serve...* has already been officially published in the country. Today, Lacoste supervises students from Brazil—which allow his performance to be verified. Have the Lacoste years come to an end or are Brazilians open for a post-Marxist, or rather, a postmodern Lacoste?

References

Azevedo A (1956) Discurso de Abertura do I Congresso Brasileiro de Geógrafos, da Associação dos Geógrafos Brasileiros, vol VIII, Tomo I, 1953/54. AGB, São Paulo

Chonchol J, Martiniere G (1985) L'Amérique latine et le latino-americanisme en France. L'Harmattan, Paris 332 p

Correia RL (1968) Os Estudos de RedesUrbanas no Brasil ate 1965. In: GEOGRAFIA URBANA, Comisión de Geografia, IPGH, No 274. Rio de Janeiro, 1968, p 192

Faissol S (Org) (1978) Urbanização e regionalização. IBGE/Secret. Planej. daPresid. daRep., Rio de Janeiro, 247 p

George P (1963) Alguns Problemas do EstudoGeografico da Populção. In: Visita de Mestres Franceses, IBGE, Rio de Janeiro, p 31

George P (1976) Annales de géographie. LXXXV:48–63

George P (1990) Le métier de geographe. Armand Colin, Paris, 250 p

Monteiro CAF (1980) A Geografia no Brasil (1934–1977): avaliação e tendências. Instituto de Geografia, São Paulo 155 p

Muller NL (1968) Evolução e estado atual dos estudos de geografia urbana no Brasil, n. 274. Instituto Pan-Americano de Geografia e História, IPGH, Rio de Janeiro

Valverde O (1989) La cooperationfrançaise dans la geographiebresilienne. In: Cardoso LC, Martiniére G (eds) France-Bresil, Vingt ans de cooperation, no 44. Collection "Travaux et Memoires". GrenobleIHEAL, PUG, Paris, p. 83 (Obra coordenada POT)

Chapter 5
New Geographers Enter the Picture

Abstract This chapter continues the discussion of periodization and from this angle deals with the arrival of new French geographers on the Brazilian scene and takes as its reference point those professionals who reinforced the French way of doing geography after the hosting of the UGI Congress of 1956, in Rio de Janeiro. It considers the affirmation of French geography after that event as fundamental, especially the activities of professionals of the stature of Pierre George, Michel Rochefort, and Bernard Kayser. Afterward, it focuses on Yves Lacoste and the great interest provoked by his ideas, increased by the way his book *A Geografia Serve antes de mais nada para fazer a Guerra* was read in Brazil. It emphasizes the success of the book in the circumstances the country was going through in 1977–1978.

Keywords Affirmation · Professional action · Conjuncture · Influence

An analysis of the French influence on our geography, identified from a set of formulations, ideas, and points of view grounded on that country's strict geographical principles, results in the personification of the professionals who passed through Brazil and, given the peculiarities of their work, were able to form groups and continue the so-called French school. The approach of the new geographers takes as a reference those professionals who reinforced the French way of doing geography in our country after the IGU Congress was held in 1956. It is from this event that we have emphasized the actions of professionals with the standing of Pierre George, Michel Rochefort, and Bernard Kayser. Subsequently, Yves Lacoste and all the uproar caused by his ideas emerged, accentuated by the interpretation that was made in Brazil of his book *A Geografia, isso serve...* brought about by the crisis that the country went through in the years 1977/1978.

Addressing the new presupposes clearly establishing what the old is. It is evident that we have not used a chronological criterion although we recognize the need for its use at various points as it enables the sequencing of events and/or episodes. At the same time, the model's approach has methodological restrictions as it imposes a link between French geography's influence in Brazil and the arrival and permanence of geographical professionals in the country, thus excluding those who have

© The Author(s) 2016   71
J.B. da Silva, *French-Brazilian Geography*, SpringerBriefs
in Latin American Studies, DOI 10.1007/978-3-319-31023-7_5

published books and supervise research but do not maintain a more effective link with the country. Mindful of this detail, in the case of this study, we have avoided excluding names, at least the most expressive who have left their mark through their style of writing, research, or supervision.

Many French scholars passed through or had relationships with Brazil, both directly and indirectly. As part of this process, we can identify the names of professors like Elisée Reclus, author of *The United States of Brazil*, edited in Brazil by Garnier, in 1900 in Rio de Janeiro and Pierre Denis, perhaps the oldest reference points of geographical science about the country. Denis wrote *Le Brésil* for the Geographie Universelle collection, which was directed by Paul Vidal de la Blache. Speaking of French scholars immediately brings to mind the names of Deffontaines or Monbeig.

However, it is worth remembering the stature of professors like Roger Dionwho substituted Monbeig. In 1948, Brazil received Pierre Gourou, the illustrious professor at the College de France, a specialist in tropical regions and author of the famous book *Les Pays Tropicaux*, edited in Paris by PUF in 1969. Another specialist is tropical countries, Louis Pay, arrived in 1950 from Bordeaux. From 1941 to 1952, Francis Ruellan, professor of applied geomorphology, was in Brazil, almost exclusively in Rio de Janeiro, going to São Paulo in 1952. He was succeeded by Andre Libault, who remained in the country between 1965 and 1973 as a professor of cartography at USP, where he gave several courses and directed the *Atlas de l'Etat de São Paulo*. In 1973, Andre Journaux arrived. Among the most recent visitors are Herve Thery, author of the book *Le Brésil* published by Masson in Paris and first edited in 1985 who has occupied the role of director of the RECLUS Group in Montpellier, an accredited researcher at CNRS and a member of CREDAL (*Centre de Recherche et de Documentation sur L'Amerique Latine*) among other activities. The names of Martine Droulers, the current director of CREDAL's *Brazil group*, organizer of the book *Le Bresila l' Aube du Troisieme Millenaire* and who taught at the Federal University of Paraiba.

Also of great importance are the names of Héléne Lamicq, Bataillon, Laserre, Gervoise, Labasse, Helene Riviered'Arc, Leloup, Pébayle, Juilliard, Calembert, Gallais, Noin, Burgel, Pinchemel Bennefont, Claude Collin Delavaud, Anne Collin Delavaud, Babinaux, Coing, Dupuy, Malezieux, Durant-Dastes, Revel-Moroz, and Fournie, among others, who in addition to their research were noteworthy for their supervision of theses about the country.

Focus is being placed on the new due to their greater influential capacity and the content of the interviews that reveal how the relationships that led us to consider Paul Claval and Jacques Lévy's performance as neware established. As seen above, the idea of new is not specifically tied to chronological age, but to the beginning of the new contact and the possibilities of polarization and group formation by these two professionals. Other professional geographers worked in the country during this period; however, when we look at the journey of each of these two and how they established their relationship with Brazil, it explains why we call this the phase of *new geographers*.

Nowadays, Claval is one of the most renowned French geographers. He has brought together a significant number of quality professionals in his laboratory and is a major promoter of geographical events in that country. His works are known worldwide; however, in Brazil, few are familiar with his output, *Space and Power* being in those days his best-known book. He is noteworthy as the thesis supervisor of Brazilians in France, and this permits his entry into the list of new geographers. Claval supervised and is supervising theses by researchers from Rio de Janeiro, Florianopolis, Belo Horizonte, and Fortaleza and maintains intensive academic contacts with university professors and researchers of the main geography departments of Brazilian universities and other research bodies. He is the author of the book, *The Construction of Brazil,* edited in Paris in 2004 by the Belin publisher.

Jacques Levy is a great interlocutor that Brazil has in France. Jacques Levy is undoubtedly an excellent example of the new teachers and researchers who have entered the scene in support of scientific relations between the two countries and is a leading figure in organizing geographers in France, having also held the vice presidency of the French Association for the Development of Geography. He maintains an intense exchange with Brazilian professors and researchers and is an obligatory figure for contact and discussions by Brazilians who are in Paris for long or short periods. It should be noted that, coincidentally, both of the teachers included in this condition of *new French geographers* have some proficiency in the Portuguese language and convey a reasonable knowledge of the country. The choice of these two does not indicate the exclusion of other teachers or researchers working in Brazil.

In the interviews carried out with both professors, we collected material and information of inestimable value for a more secure and deeper analysis of the relationship between the two countries.

Paul Claval, a distinguished researcher at the University of Paris IV, an institution where his activities and permanence guarantee his unequaled prominence and respectability, is one of the main exponents of contemporary French geography. *Space and Power* is his best-known work among Brazilians. He also published in Portugal the book *The New Geography*, published by the Livraria Almedina of Coimbra, Portugal, in 1987.

With his vast intellectual production, without a doubt Claval is currently one of the most sought-after teachers in France and abroad. His work has been translated into several languages. His doctoral students come from all over the world, with an expressive number of Europeans and oriental students, which is not common among students enrolled on doctoral geography courses. As a visiting professor in foreign universities, he has made his contribution in Canada, the USA, Brazil (he gave a course in Portuguese at the Federal University of Bahia), and China (Taiwan), among others.

The number of Brazilians who seek him out for thesis supervision is increasing. Claval has been the director of research at the Space and Culture Laboratory, which has its own publication and a dynamic group of researchers who present their results at regular seminars promoted by the laboratory that has become a privileged space for the discussion of the themes presented. In its vast, rich, and estimable intellectual production are topics linked to its preferred themes: the history of geography, *amenagement*, urbanism, political geography, and cultural geography. Claval's doctoral thesis was supervised by M. Chevalier and defended in Besançon in 1970. Among his published research and books, the following can be highlighted:

- *La région nouvelle a la fin du xx érne siécle* In: Treballs de la Societat Catalana de Geografia, 1989, no 17, pp. 1109–112.
- *Les geographes français et le monde méditerranéen*. In: Annales de Geographie, 1988, vol. 97, no 542, pp. 385–403.
- *Le théme regional dans la litteraturefrançaise*. In: L'Espace Geographique, 1987, Vol. 16, no 1, pp. 60–73.
- A critical review of the centre-periphery model as applied in a global context. In: The World Economy and the Spatial Organization of Power/Shachar A., A., Orbeg S., eds (Avebury: Aldershot, 1990, pp 13–27).
- *Criativite Culturelle et grandes capitales*. In. La geographie de la criativiteet de l'innovation/Chevalier .., dir. (Paris: Publications de l'Universite Paris-Sorbonne, 1990, pp 53–63).
- *La conquete de l'espaceamericain, du Mayflower a Disneyworld*. (Paris, Flamarion, 1990, coll. Geographes., 320 p.)

In addition to these works, Paul Claval wrote *Les mythesfondateurs des sciences sociales*, *La logique des villes*, *Principes de geographiesociale,* etc. On the theme of the renovation of geography, Claval affirms:

> Classical geography allows us to describe and understand the rural environment, the realities of regions or the ancient provinces. Industry, the city, tourism, the migration of populations and the fearful rhythms of advanced society are beyond its scope.[1]

By pointing to classical geography's inability to perceive the various faces of modern life for analytical purposes, Claval indicates that there has been progress; however, he is more moderate and parsimonious in recognizing that geography did not build this path alone. The author is emphatic;

> The transformation required is very advanced. The renovation was the work of geographers but on the same level it was also of sociologists, economists, ethnologists or urbanists. Historians participated in the movement, but their contribution was less essential than in the former period.[2]

According to Claval, the renovation began to be felt from the 1960s; the author states that there was a certain hesitation as to how it was categorized:

> Some talk about theoretical geography, others of a quantitative revolution: these are expressions that, whilst not inexact, only express half of the reality. The new geography was born in a period of intense intellectual ferment; it developed in an atmosphere of social agitation. At last Marxism, which hitherto had only played a secondary role in geographical

---

[1]Claval (1987, p.9).
[2]idem.

thought, took an interest in these developments. Young theorists clamored for the opening, through an epistemological break in the manner of Althusser, of the geographical continent of scientific knowledge.[3]

This text reveals that the author treats the renovation in a comprehensive manner, taking into consideration all the lines of thought with different colors and sources in accordance with their political–ideological options.

5.1 Interview with Paul Claval

I ask Paul Claval about his opinion on Brazilian geography and whether it has its own profile and constitutes an autonomous school. He replies as follows:

> For me, in principle, a school of geography presupposes a large group of professionals, and a body of doctrines that are transmitted from one generation to the next. On the other hand, in the strongest sense, they are a group of professionals who develop original conceptions of the discipline, in a strong and a weak sense. In the weak sense they are a group of professionals with continuity, which applies perfectly to Brazil. Nowadays, after 60 years, Brazil has a big group of geographers, with well qualified people, with tradition, that has been consolidated and in this sense it is realistic to talk about a Brazilian geography. It is a school in the same way as Iberian or Italian geography. I would say that in my view, considering what happens with Iberian geography, with Italian geography, it becomes more difficult to apply the concept to Brazilian geography because it has not developed a completely original body of doctrine, but it is possible.

> I think, in the history of geography, there was a period from the late nineteenth century to the 1950s, when in fact general human geography adopted currents, very different parties, very contrasting views of geography and so that at that time, effectively there were national schools. There was a French school, a German school, an American school and, at that time, an English school. Therefore, there were geographers in England who had professional level with a series of traditions emerging. And there was a Spanish school taking shape, and there was a Polish school. I believe, consequently, that the idea of an autonomous school is something that applies to a particular time in the development of geography in the world. I am sure that currently there is a strong opposition between French geography and Anglo-Saxon geography, somewhat different habits, but I believe that this idea comes from the "national", in the strong sense of the term, and does not apply to the current reality. In this sense I think that the fact that Brazilian geography is very equipped is proof of its maturity, and I think, characteristic of the whole of geography in the world today.

I continue asking Claval about the phases of hegemony, of distancing, of rifts, and of drawing closer in the scientific relationship between France and Brazil in the field of geography. I mention the direct and indirect presence of various professors and cite Deffontaines, Monbeig, Papy, De Martonne, Pierre George, Kayser, Labasse, Lacoste, Tricart, Pebaylle, and himself, among others, who are constantly being cited in our courses and works. I ask whether he thinks that there is a discontinuity in our relationship and any symptoms of stagnation.

---

[3]idem pp. 11–12.

I think there have been highs and lows in the relationship between Brazilian and French geographers. The same occurs between Brazilian geography and French geography. There have been periods of very intense relationships, when the French geographers conserved the German *college* model and there was the propelling role in the birth of Brazilian geography with Deffontaines, Monbeig, Papy, Pierre George, Rochefort, Kayser and others, revealing a period that went up to the end of the Second World War. I think from that moment some Brazilian geographers had a desire for independence which, I would say, is very legitimate, and that does not translate in Brazilian researchers in the search for other models; a typical case is that of Sternberg, who broke away and finally settled in Berkeley. It seems to me that there is a generation that turned to the United States and in Sternberg's time it was not the case of a transformation in the conception of geography; this was more typical of the 1960s with the *new geography* of quantitative methods. There was, for some Brazilian geographers, the impression that what came from the old continent was outdated and it was worth turning resolutely to the United States.

The left-wing orientation of certain Brazilian geographers evidently limited the role of people like P. George, B. Kayser and M. Rochefort retaining the importance of the 1950s and 1960s. But I believe that there was a rapprochement rescheduled from the early eighties and I think there are two phenomena. On the one hand, the discovery that quantitative geography, the new geography, has limits and, consequently North America could only offer a partial model; on the other hand there was the emergence of new guidelines connected to the modernization of French geography, this is not the regional geography of the 1930s and 1940s, but a Marxist leaning geography that Kayser, Rochefort and P George could represent in the 1950s and 1960s, which are increasingly broader sources nowadays. Beyond the interest there is in France for the problems of spatial organization, in *amenagement*, of geography applied to one part and the problems of historical and cultural geography, among others. And after a renovation, it can only be said that France has reencountered its original influence, more selective now, able to advance in the discipline. What I think is that there is a moment of resurgence.

I ask the professor whether he can trace his intellectual itinerary referring to the role that the work and contact with Brazil has had on its formation.

In the countries where I have been at different times in my life I can look back on my stay as a *petit français* in the United States, New Zealand and Brazil. As a result, and for very diverse reasons, this only comes from the countries that I visited very early, and I learnt about Brazil when I was a student in Toulouse, from young Brazilian colleagues who were preparing their theses. One colleague was a geographer and her husband was a doctor. I believed I could get to know Brazilian geography and I found that although I did not know Portuguese, I could read Portuguese, and I could write in a journal and so at that time, there was a renewed interest in Brazil. I can say that the works we had about Brazil interested me very much because there was a different society that interested me because I had not found in French literature, the equivalent of what Pierre Dennis, who also lived in Brazil, had written about Argentina. It is a magnificent geography, one of the finest regional studies. Dennis worked in the archives in Argentina before the First World War and he worked the same way in Brazil and, consequently, I had the impression of a work that at certain times I used as a traveler's testimony.

Next, I single out the habit of presenting the geography of Brazil through the history of sugar. It is interesting, but at the same time, from the 1950s French geographers repeat themselves and they had the view of sugar in the Northeast, in Minas Gerais, in Rio, in São Paulo, a succession of development bases. The Amazon later and I noticed that something was missing, that after having told this cycle I would not know what to say about the periods of the country's history, of its organization. I had the pure impression of a resurgence which lacked substance. The one that always seemed different was Monbeig's study that rightly, in the case of São Paulo, gave a much more coherent vision. It was reading Monbeig that I discovered an interest in medical or health geography and it was fun because I was a student when I read Monbeig, I had never seen, from a precise concrete example, the approach of medical geography.

What I found interesting in Pébaylle's approach is that the work is an exact reflection on the ordinary conditions of spatial organization in Brazil, where it is not only the history of these pioneering fronts that is addressed. There is a reflection on the specificity of Brazilian space, a reflection that, deep down, was not found in French geographic literature. It gave me the impression that he focused himself after Monbeig. For nearly thirty years the discourse on Brazil was repetitive, whether about pioneer fronts or the sugar cycle. They are minor works, we always find the same view of sugar, of the pioneer fronts and many other things are lacking for an understanding of Brazil. At that moment, I had the feeling that Brazil had been poorly studied and I became interested, when still a student, in going to Brazil, as a cry of independence, and I had a lot of things at that time, but I tried and always kept the memory of a country about which something was missing in French-language literature.

A little remained, on the surface I think there was a historically based interpretation but that was distant from the Brazilian historians. This interpretation did not resemble anything by French historians. It was only avoided by geographers, who sought to understand what the Brazilian space is in itself, as the pioneering fronts have passed and society organizes itself as it functions, how social competition is organized, how the issues of the key-roles in Brazilian society and the formation of elites are addressed. These are essential things to understand a country and I did not find them in the works I consulted.

I ask concerning the rapprochement of French geography with Brazilian geography and vice versa—because I think he is a representative of this rapprochement. I think the first person Paul Claval supervised was Leila Cristina and then Margareth Pimenta, Paulo Cesar da Costa Gomes, Sergio Augusto, Maria Clélia Lustosa da Costa, Maria Geralda, and myself are doing a postdoctoral internship. In this instance, do all these theses indicate something new in the treatment of Brazil, a different approach to that normally found in France? Paul Claval answered as follows:

I think that though the work of the Brazilians I have directed there is an idea of modernity, a condition of the formation of a modern urbanized space. A geography of communication in the interior of a unified Brazil, that does not circulate well, that presents itself in Margareth Pimenta when she studies the condition of the textile industry, and more recently in the work of Leila Christina on the role of telecommunications in forming the functioning of the banking network in today's Brazil. From this point of view, the Brazilians have discovered the importance of telecommunications during their internships in France and among the ideas that seem important to me is the role of the elites in the spatial functioning of a society and something that is above the level that has attracted the attention of my Brazilian students and that I believe is important in understanding a country's history. For me, the elites more than the classes, because a lot has been said about class and region but it seems interesting to see how the country's responsibility emerges, to strengthen the knowledge that allows it to be organized and this constitutes a network through which information circulates and begins to penetrate in the same way that it crosses the notion of the specificity of space. I believe that there is still a lot of work to do in Brazil by Brazilians to try and go further this time.

I used the opportunity to ask what Claval thinks about the practice of French geography regarding it if it is similar to colonial geography or tropical geography. In response, Paul Claval was emphatic:

I do not think it is similar for various reasons. The first reason, I think P. Monbeig had a very different preoccupation than geographers like Guorou and Pelissier. It was different because he obtained more strictly geographical witnesses than the others, because he had a curiosity that was not limited to physical geography, limited to the traditional fact of agriculture or land use and I believe that he quickly understood the importance of circulation, the importance of history, the role of the settlement system. And he brought a vision of society from the study of Brazilian society that was not present among experts from the world of colonial geography. The action that they had on the social facts in colonial geography, does not apply in the case of very traditional structures because from the beginning, what interests Monbeig is the role of São Paulo, the role of the metropolis, the role of the railways, the role of a certain innovative capitalism. I believe that this is not a simple equivalence to the rest of colonial geography, the rest of tropical geography, but he takes advantage of tropical geography, for example, all the research on medical geography, and integrates it in his approach. From this point of view, he is part of the tradition of analysis, but it is not exclusive. So originality is stronger in Monbeig than it was in the others. I believe that other French geographers accepted this interpretation of the history of Brazil in terms of sugar more easily. From this point of view, it was little different from the approaches traditionally practiced by geographers but it became repetitive. He accepted an interpretation without seeing if it covered Brazilian reality and would allow him to understand the dimensions of the progress of the settlement that is one of the elements of regional diversity. But certainly there were other things to consider in Brazil at that time, which were different; although exceptional they were not the same as the ones understood by Monbeig or even by other French geographers who worked in Brazil.

I ask whether there is anything new in French geography, and in the case of a positive answer that he could tell me what it consists of what are its characteristics and key personalities. Polite as always Claval gave his reply, "You have asked me an extremely difficult question because you know the subject more than me and then I get a little shy about answering this question. What's new in France, after so much time? Please specify." I affirm that I am referring to the current moment (1992) and that when "we Brazilians are discussing this, we have noticed a sort of *loss* of Paris and the resurgence of other so-called traditional centers for Brazilian geography like Strasbourg, Toulouse, Bordeaux and Caen, but we wonder, for geography in general today, what are Montpellier, Grenoble, and universities in Paris working on with geography, such as Paris I, N Paris, Paris VII Paris VIII, Paris XII." Claval responds:

I also think that French geography at the present time has diversified its interests and has renewed itself in the concept of the region, it started with the Reclus Group in Montpellier, and the reflection about the region was partly expressed by the works of the Reclus group, by a systemic effort to use the map to express this. The idea of analyzing the country in the form of a simplified geometry is not always unanimously accepted, but it is something that, from a pedagogical point of view, is an important advantage, and Montpellier's Geography House offers researchers who wish to get documentation of France and Europe informative and modern cartographic tools and first-rate facilities. Therefore, the *Maison de la Géographie* of Montpellier is an important center in areas that had been neglected such as political geography which has become an important component after ten years. From this point of view, there are interesting currents of reflection on political geography and geopolitics. It may be the level of research at the University of St. Dennis-Paris VIII - around Yves Lacoste where training is better structured for the third cycle in this field, doctoral preparation. And there is interest in social geography, an interest that developed after the 1980s around geographers at the University of Paris I, geographers at Lyon II, and geographers from the University of Caen and other western universities - Nantes, and minor centers. There is very interesting research at the limit of demographics, wealth, poverty, exclusion phenomena; research projects are very numerous in this group. After ten years, also we are witnessing the renewal of historical geography and cultural geography. Historical geography and cultural geography have always been practiced, but were considered somewhat marginal after the 1960s and especially the 1970s. Currently they are active subjects for young people and in matters of historical geography in centers that specialized themselves. The same is practiced in several universities but with regard to Cultural geography it is the Paris IV that maintains it. At the time the research was carried out the most important research center was there with an interest in the history of the landscape. The work and research of M. Pite, reveal a strong interest in landscape understood as being able to explain reality. The same happens through the work of the Contemporary Japan Research Center with Augustin Berque. It is in Paris that the best and essential centers are found. If I leave the field of human geography, social geography, to see what is happening in the field of physical geography or natural geography. The impression is that we are experiencing a transitional phase and after the 1950s, 1960s and 1970s, we are witnessing increasing specialization even if we are more interested in the overall approach. It gives the impression that physical geography becomes increasingly like specialized geomorphology, even if it is not said, a biogeography punctuated by climatology, which has considerable encounters in dynamic meteorology, the study of types of weather, of air masses, slightly different movements, and gives the impression that this very specialized physical geography turns into a physical geography for man that interests us, above all, the functioning of ecosystems in which man is involved, the environment and its preservation, interests us, the weather, which means it is from this point of view, this rapid mutation somewhat explains the overshadowing of traditional centers that have not disappeared, but are of interest above all, today, to the geologist and the specialized geomorphologist, who are not geographers. But there are interesting studies that continue to be done in Strasbourg, Caen and Grenoble, which is a traditional center of this specialized geography, and there are new guidelines addressing the environment, as in Nice, for example, Grenoble and Nanterre. And there are works that surpass in depth the approaches of Physical geography. The regional field of guidance, I would say, remains interested in the tropical world with less success after a few years, but the Mediterranean universities continue working on the Maghreb, the Mediterranean world and the Middle East.

Toulouse is directed toward the Iberian world especially Hispanic American and Spain but it also has a clear interest in the Spanish-speaking world. And what changes up to the present are the research centers on Europe, which seems paradoxical; centers working in the field have a very clear interest in the current transformation in the East, half of the pacific world. The Japan Study Center Contemporary participates in this but there is also a lack of more significant study centers on Eastern/Central Europe, the East, study centers on the Far East. I believe that there are specialists, scholars, students, but there are no centers that are up to addressing the problems facing these countries.

Continuing the professor starts to speak about Brazilian geography:

In my view, Brazilian geography has become an adult geography with an abundant production of quality journals in which there are, at the same time, articles, analyses by Brazilian authors, analyses of Brazilian works and there is an effort to keep up with all the research worldwide. I think, for example, that the role of Christofoletti's journal, the *Revista Teorética* has paramount importance in this area. I have the impression that through the texts and the abundant and numerous works that this magazine publishers, it is possible for Brazil to follow what was done in the Anglo-Saxon world and Europe. I believe that Brazilian geography has come of age with a list of professionals capable of dealing with national issues and keeping abreast of interests and what is happening on an international scale. There are secondary and tertiary level textbooks that are well written and have been published for many years. They are annuals equivalent to those found elsewhere in the Western world. There are translations and it is possible to find English works, French and many other existing works in Portuguese, there are important fundamentals and statistical demonstration studies. I believe what hinders Brazilian geographers at the present time is something that hinders all intellectual life in Brazil, the difficulties of life in a bad system, in which teachers always have to perform a difficult exercise to finish their work. From this point of view, what limits the productivity of geographers in Brazil is that they are forced to give other courses, to accept other diverse responsibilities and, this does not correspond to a willingness to participate in the general movement of economic life and in other phenomena in the country.

The relationships I have had with Brazilian colleagues have always been interesting relationships for me, and I have the impression of being in a land where one talks to equals, talking to people who have done everything to keep up to date, to maintain their level and from this point of view, I have had very good impression of Brazilian geography, though, I have no illusion that does not have any weakness, that Brazilian geographers are in the fore. No! In every large scientific community there is diversification in terms of effort, and the results obtained.

Unlike the other geographers, Paul Claval's answers are structured in such a way that nothing goes unanswered. The interviewee was at ease and gave his opinion in response to each request, revealing a profound knowledge of Brazil. According to the declaration, his interest in the country dates back to the period when he was a university student in Toulouse.

Prof. Claval masterfully records all the spectacular works by French scholars who analyzed Brazil, such as P. Dennis and P. Monbeig, among others. His analytical ability puts him in a prominent position among other French masters who maintain ties to the country, although he did not register an awareness of work by Brazilian intellectuals on the various interpretations of Brazil, whether by Sergio Buarque de Holanda, Caio Prado Junior, Gilberto Freire, and Celso Furtado, contemplating the most varied theoretical nuances. He cites the works of his supervisees who undoubtedly conveyed a picture of Brazil and its entirety. Claval is also another very interesting case of a doctoral advisor. Despite his old bond with the country and his knowledge of the Portuguese language, he has only recently started his orientation activity. It is noteworthy how Brazilians seek out Claval either for long-term guidance like doctoral theses, or for contacts, interviews etc. His friendliness and good will make him one of the most significant French geographical characters for Brazil today.

Another geographer presented as *new* in the relations established between France and Brazil in geography is Jacques Levy.

Jacques Levy has emerged as one of the great revelations of French geography, taking on increasingly prominent positions, with a varied and high caliber output. His engagement with Brazilian geographers, his interest in our geographical production, and his knowledge of Portuguese mean he is constantly sought out by Brazilians studying in France or those heading there for research and contacts. All this qualifies him among the new geographers who are entering the scene, and his formulations open many possibilities for a meaningful academic exchange between Brazil and France. In his interview, Jacques Levy strikes a new note, exposing the existing relationships within the organization of French geography without camouflaging their ills, the power struggles, and the organization of associations.

Born in 1952, a former student of the Superior Normal School of Cachan, in greater Paris, later Jacques Levy took a position at CNRS and was in charge of research at the Strates laboratory at the University 7: C, Paris I, and Sorbonne. At the time of the interview, he was still the *maitre des Conferences* (a career level in French teaching) at the Institute of Political Studies of Paris. On the January 4, 1993, he defended his state doctoral thesis (*Doctoratd'Etat*) entitled *L'espace Legitime* which had Milton Santos as a member of the examination board that also included Paul Claval, Jean Bernard Racine, Olivier Dollfus, and Remy Knafou. Cofounder and animator of the journal *Espaces Temps*, Jacques Levy brings together a set of significant experiences in geography.

The book *Le Monde: Espaces et Systemes* organized by Jacques Levy, Marie Françoise Durand, and Denis Retaille was launched in 1992 by the publishers Presse de la Fondation Nationale des Sciences Politiques & Dalloz. It is a dense work which discusses rich and varied themes such as "Around the model-State" and "Towards a society-world" by various authors. The book includes a text by Milton Santos entitled "São Paulo—a center in the periphery."

In another work from 1991, organized by Jaques Levy, *Geographies du Politique*, in the series *Referencias* edited by the *Presses de la Fondation Nationale des Sciences Politiques*, the authors brought together a quality team and was highly praised; the newspaper *Liberation* wrote:

> How does political function articulate with other dimensions in the social space? This work aims to answer that question. To begin with the analysis of the *donne* of different types of political space; followed by the *mises* of disciplines, such as political science and geography that invested in the field: sociology, anthropology, demography, history. Once these are outlined, *les cartes* verifies the adequacy of the methodologies to regions and historical periods (Marseille, Bretagne, the USSR, and Africa). One last item on the *jeu* tries to give some reference points for thinking about the state and the movement of today's world.[4]

The famous *Le Monde* newspaper commented:

> This book is a collective reflection on the theme of political space. Specialists from various disciplines (geography, politics, anthropology, sociology, history, and urbanism) participated in the book, revealing, as Jacques Levy noted, a *degeopolitization* of the world and a dislocation of the political in favor of an extension and a differentiation of their spaces. The whole constitutes a rich and dense dossier.

The two reviews from famous and respected French newspapers were transcribed from the book's cover.

---

[4]Berque (1992, p. 358).

5.2 Interview with Jacques Levy

Jacques Levy was very welcoming when I requested the interview. At his home, after presenting the reasons for the research, I asked him whether Brazilian geography already had its own profile and up to which point it formed an autonomous school. The professor answered as follows:

> Well, before answering that question I want to make something clear. I am not a specialist on Brazil and I am not a specialist in the history of geography. Therefore I am not a specialized expert to be able to answer the question in depth. It is more of an impression that I can give of two very short visits I made to Brazil with a ten year interval – 1982 and 1992 – and my personal knowledge I could have with some Brazilian geographers. I have this impression and can certainly answer yes to this question, because compared with other countries; my feelings are that geography in Brazil is important, dense, structured, and complex. There are many schools of thought inside geography, but in any case, it resembles geography in those countries where the science is strong. I did not build a strong knowledge about this because there is a big crisis in countries in the developed world, such as Great Britain, France and Italy. In Spain it is a little less. And then the United States that is not very powerful except in a few isolated places, often behind the Europeans. In Japan there are also geographers but I'm not sure that there is a real Japanese school of geography. And Sweden must also be mentioned. So I think that Brazil, from this point of view, ranks among the large countries of geography, notably as regards Latin America. I believe that only Mexico could somehow rival, but anyway I do not have knowledge of this, and I know that there are geographers in all countries. There are in Argentina, Venezuela, and Peru, but in Brazil this context is the most important, because if there are not many geographers, it is difficult to have geography. From this point of view, in terms of mass, Brazil meets the conditions. In terms of ideas, I believe your strength is being connected to the outside, not a province, as they say in France, that is, an enclave, barely linked to the outside world, living its own pace, which could be said about French geography in the past. But we know that Brazilian geography is not *provincial*, it is connected to the flow, the intentional market of ideas. Brazilian geographers read what is being done abroad, they have a good Anglophone and Francophone culture, and produce Anglophone and Francophone work, they are important as a mass. So they are a group that is open to the world. Basically, truly, I only know a small number of geographers, but I think there is a whole part I do not know completely, for example, those who are connected to the IBGE – who work with quantitative geography based on statistics, positivist, if said in a derogatory manner. This is a geography that comes from the moments of glory of the *Brazilian miracle*, very economist-based, which has not disappeared completely. I do not know that it exists or existed but I do not know it. What I know is the geography that is considered a social science, which thinks seriously about the concept of society, precisely the social totality, which is the idea that space has a dimension of society. Above all, it uses qualitative methods more than quantitative methods because it thinks they are the best way to grasp its object. All these things link it to comparable schools in Europe. As I am part of this approach, evidently I was in contact with this version, this part of Brazilian geography. Afterwards, my knowledge linked me strongly to Milton Santos, who acted as my ambassador in this access; so from a certain angle, to be honest; I should say the Brazilian geography I know is the one that gravitates around Milton Santos.

I interrupt and provoke Jacques Levy once more asking whether Brazilian geography would be a type of school and the professor continues:

Yes, a type of school, especially now that Milton Santos is at USP, where a lot of people are working – working well and a lot! – I think this reinforces when I compare it to the time I was in Rio de Janeiro. I also met a third aspect of Brazilian geography that is the aspect of militant geography, during my first trip in 1982 to the AGB Congress in Porto Alegre. I was struck by two things I had not seen elsewhere, that is, the possibility of getting two thousand geographers participating in a place that was far away for everyone, and expensive, because transport is expensive for Brazilians, and among the participants many secondary school teachers. This seemed to me a very important sign of vitality. I believe it was the same in Presidente Prudente.

I inform him that the AGB has always managed to attract a considerable number of participants to its events. Jacques Levy continues:

And the first aspect is the fact that students are in contact with users of geographical thought, especially teachers. Another aspect of this, more cyclical and probably more linked to the time, is the first step of the political struggle for democratization, and certainly there was some mix of genres, I would say, between strictly scientific work and political activism. It was both pleasant and disturbing because it was thought that scientific autonomy risked being diminished by addressing the political problems that were evidently many and important, and I understood 'why'.

I alerted the professor that in the context of the event in Porto Alegre, one could already observe a schism, a separation. Jacques Levy agrees and continues:

Effectively that was my impression. After ten years what amazed me was that Brazilian geography works, works hard and has theoretical ambitions while working in the field. So it has, I believe, a good balance between these two aspects of the research.

I continued provoking Levy stating that Brazilian geography maintains strong ties with French geography, presenting differentiated phases such as hegemony, withdrawal, rifts, and rapprochement in our scientific relations. I mentioned the name of famous teachers of the level of Deffontaines, Monbeig, Papy, De Martonne, Pierre George, Kayser, Rochefort, Labasse, Claval, Lacoste, Tricart, Pebaylle, Levy himself, and many others, who are constantly cited in our courses and our work. From this preamble, I asked him how he saw this context, if there had been a discontinuity in these relations or if there were symptoms of stagnation. Jacques Levy raised the question "When you say symptoms of stagnation, what is stagnating, the relationships?" I say yes, that the relationships which you referred to in the previous question, including the relationships between Brazil and the United States, quantitative geography and so forth. Jacques Levy answers as follows:

I have the impression of a certain current asymmetry as Brazilians are very interested in geography from other countries, especially French geography. There is continuity, from the Brazilian's point of view, because of the interest that they have in France. Inversely, I have the impression that the continuity is not that strong because the French like to use Brazil as an object/field for research, and eventually as a public, but I believe it does not show much interest, there are brilliant exceptions. Many cannot accept the idea that Brazil also produces geography and that they should also go to Brazil in the same way they go to Great Britain or the United States to enrich themselves at the source of production. But I believe they also do not know Brazilian geography as a producer of works. The Third World finally became popular in our country in the 1960s and 1970s, and more so today, and something that widely permeates geography is the problem of the ultimately political and social hopes that many intellectuals put on Third World. They had the impression that the periphery was finally the hope of the center. Today, the dominant idea is that the center is the hope of the periphery.

Well, there is, no doubt, a lot of excess in the colors. I think that is the ideas that dominate today, that is, we seek the answers to our questions precisely in the South. So, there is probably a loss of interest when we see in the countries of Latin America and particularly in the Third World, the inconvenience of not being the countries that have recently left the Third World such as those in Southeast Asia, or countries that are in abject poverty as those from Africa. Therefore they are more complex and probably are not open to a simplistic reduction. It may be that this is no longer the interest of those who are researching with simplistic worldviews. I think that is why there is particularly an eclipse of Latin America as the center of interest. Emphasizing, I believe there is even a continuity that is linked to the fact of French geography not doing very well at the moment, at least institutionally. This is in the face of the regime in the 1970s, when due to internal crises in geography and because people were more interested in themselves than in the outside world, and all this to solve the problem of defining geography. Nowadays geographers show some improvement in diversity. They improved, because they are linked to the exterior and so are more connected to the universal geography. Those of Reclus, the festival of *Saint-Die de Vosges*, are a kind of desire for a new universalism that inevitably caused contacts with the outside world, but a deficit of relationships remains, because in the end, you are always listening to French geographers but French geographers should also listen to Brazilian geographers.

It was hard not to demonstrate my enthusiasm with the answers and continue the interview enquiring the professor. I refer to his visits to Brazil saying "you said that you have been to Brazil twice and that you know many Brazilian geographers. You spoke of Milton Santos, Maria Adélia, Ana Fani, and Rogério Costa. Can you trace your intellectual itinerary making reference to the role that work and contact with Brazil played in your training?" Always attentive and careful, Jacques Levy answered thus:

Yes. It's hard to take stock when you are quite involved in something, and Brazil's place was reinforced in my life. So it is difficult to take stock. There are roughly two phases in my personal Brazilian history. The stage at which I essentially met Milton Santos and we exchanged ideas, he wrote articles and I went to the AGB Congress in 1982. We discussed a lot, did conferences, but these were relatively one-off events. After a few years, what was new was that little by little I found myself included in a small network of relationships with Brazilians coming to France. It is not only by going to Brazil that I make contact with Brazilians, but also remaining here. So, I had occasion to discuss and advise Brazilian students, researchers and teachers coming to France and I felt a very strong demand from them, possibly because the welcome here is not very good for foreigners. It is tough, there is coolness, and there is no ease, relaxation, which makes foreigners feel lost. I believe, among other things, that I'm a little more available and so the demand grew insofar as the flow of Brazilians coming to France increased; I mean on the whole. I cite the specific case of the experience I had with Rogério Costa because we were truly together for many months and established exchanges on a regular basis. We saw each other constantly and he participated in seminars that I organized, he intervened at different times and we discussed explanatory models of the contemporary world, the epistemology of the social sciences, modernity and post-modernity. I realized he gave me a lot while I could pass something to him. In this case there was a truly productive exchange, which ultimately may be easier among people who are in different institutions because there is a phenomenon of the sociology of institutions that means the more I'm around someone the more difficult it becomes to exchange with them, because there is the phenomenon of rivalry or feeling that the other will not add anything. Finally, exoticism as a starting point for a real exchange because you want to know the world of an individual who comes from afar. And for all these reasons I have noticed that in general contacts are easier with people who come from outside and even from far away. For this reason, I experienced the importance of the Brazilian geographers coming to France with whom I could discuss, such as Ana Fani A. Carlos, only once in France but then we exchanged things - she sent me an article, her book about the city which I found very interesting, because it has concerns that I also have about the city. So, talking about Brazil was a fundamental objective for her and in a way, also for me. This changed a lot when I went to Brazil in September and we imagined doing things together, etc. So, some things fit together, one of them was an idea, which I do not know will come to anything, about working on research on the Brazilian political space, notably the electoral space. This was a suggestion made to students and researchers at USP, but there is a political geography laboratory I cannot remember the exact name, group, or seminar. I know that it is led by Maria Adélia, a group that is working on geopolitics, on the interpretation, for example, of electoral results, about the virtues of the candidates, the relationship between politics and space; now that democratization allows you to have a mass of facts, one can discuss citizenship, in short, the meaning given to that fact. And there are many facts to be discussed, and I felt, on the other hand, some hesitation because I understand certain impressions are linked to the fact that the elections are fully free and transparent, etc., but I can assure you there is a need to know and interpret if elections are completely free and transparent.

I continue giving another direction to the research: "You even made a map with the election results from France for Maastricht and the European Union, you made the comparison."

Yes, it's a thing I did some time ago and, indeed, it must be interpreted with caution, wanting to know what you want to know, but it tells me things; in Brazil too, the results of the last municipal elections, because, for example Salvador has a leftwing mayor and on the other side, the state government, a well-known big shot. Here's a geographic issue about politics that, in my view, it is worth considering. But I'm eager to go beyond this in my exchanges and work together.

I interjected asking whether this included Fortaleza and Jacques Levy replied "This is it, I was invited in January but unfortunately I could not go. It's hard for a foreigner who does not speak Portuguese well; I used to call Salvador "Salvateur" (laughs)."

I continue and provoke the professor asking whether he thinks that the practices of French geography in relation to Brazil are similar to colonial geography or tropical geography. Jacques Levy is emphatic in denying this.

No, I do not think it is the same thing; the issue is not the same. I think there are two traditions in French geography that meet in tropical, colonial, traditional geography that are not found in the geography of Brazil done by French scholars. By tradition, is not exactly colonial geography, which is one in which geographers are at the service of settlers, the colonizers, and the French State, wanting to know the country's resources, wondering where to install the infrastructure, wondering how to control the population's movements, political ideas, and prevent anti-colonialist subversion. This whole settler and colonizer issue was more or less put into practice by geographers in their work, particularly in the French colonies. There were similar things in the British Empire, but this did not happen in Brazil.

This colonial geography has another tradition that was, I would say, an archaic tradition, which was more interested in the countryside than in the city, more by agriculture than by industry, more by industry than by services, it was not attracted to the more modern things of social life. It is also interested in what it makes possible, for its appearance, a naturalist explanatory model and the social environment, and society is ultimately what matters to geographers who attempt to apply Vidal de la Blache's concept, which means leaving permanence and not what changes. What changes, changes on the surface but the essential things are stable, they do not change and that is our work.

Hence the temptation to conserve nature because nature does not change, it is our pace that changes, so it is a relationship with nature in a fundamentally naturalist-possibilist model. It is a variant that needs introducing to the geographer's work. So there is obviously this position, this attitude. It manifested itself more in some Third World countries, in countries where the development of production is less important and where agriculture plays a more important role with the physical contrasts of the earth and climate. There is this logic that French geographers are particularly interested in Third World countries, in tropical geography, as you say in the questionnaire; tropical but not colonial. In this sense, the distinction between these two expressions, contains tradition, but we can understand the tradition of tropical geography by the fact that in France geographers can study cities because they are interested in the relationship between cities and the countryside; for example metropolitan space is barely worked on. Fortunately sociologists are interested in this. If not, we would not know much about inter-urban space and this vision. This approach, this *demarcation* that was present in physical geography is what actually made a country like Brazil, with its empty spaces, its town-spaces, its pioneering spaces and the strong presence of agriculture and a weak industrial presence, particularly attractive to geographers. They were also interested in Canada and as the world expanded rapidly, it also expanded into tropical zones belonging to the British Empire, where British colonial geographers and others worked. Brazil and Latin America in general were a space where they could go, it was not closed.

I continue by asking Jacques Levy if there is anything new in geography in France and if there is where it is being applied and who its key people are. With much attention and calm, he answered:

It is hard to summarize it in a few phrases, especially because Brazilians know French geography well and nothing escapes their notice, in this aspect. But the difficulty at the moment is due to a lack of ruptures; the most spectacular evolutions, the most visible, occurred between 15 and 20 years ago in the 1960s, when it was easier to classify the lines of thought, the ideas. Now it is more difficult because the scientific work is not concrete research that makes privileged date unviable, with nebulous ideas, schools of thought in less structured groups. For example, I think that the neo-positivist-quantitative school was never very concrete in France because it was tempered by other theoretical principles and the vision it had of things was clan based. Nowadays, quantitative studies are increasingly used, in particular the automatic management of variables, qualitative studies of relations, studies that safely use quantitative private data but take longer to interpret and manipulate the data. Of course there is an evolution, but it is not easy to say where this ends, where this starts, but there are nuances, things that evolve very slowly and at some stage, it is clear that there is evolution. As an example, I can point to an evolution in French geography, the fact that two years ago the structure at the CNRS was divided into sections that changed a lot for geography, for the first time in the history of geography. Now there is human geography that is separate from physical geography, and human geography is associated with urbanism in the section called geography, urban planning and *amenagement,* while the equivalent divisions at the university still have one system: there is human geography, physical, human and economic, the same names as in the 1950s, or even earlier. It is noteworthy that this change was not proposed by the Geographers; it was the CNRS's administration that considered it evident that it was worth grouping together the people that work together and carry out a more formally based regrouping. A geographer working on the city will never have anything to discuss with a geomorphologist but always have something to discuss with an urbanist, an urban sociologist. So it was a change guided by common sense at an administrative level but what is interesting is that it went well, it did not cause fears, tears, which is to say, there was an evolution and I could see that later.

There are things that evolve and I think that when it was said that geography was a social science, twenty years ago, I remember, I said that. I started my intervention on the epistemological level with a vision of geography as a social science, as the essence of the science of space and society and understanding the spatial dimension of society. To say this was considered a provocation and at one point I said, I will speak because anyway, I agree and I realized that there are inflexible people. I have the impression that this fight does not make sense nowadays because everyone holds more or less the same position and there are things that evolve, I would say, globally and there is a tendency for geography to be closed in on itself. It tended to be closed within itself, now it is open to other disciplines, in particular the social sciences. Evidently, geographers are finding more logic, and they feel it more necessary to read works in social sciences and even philosophy. They quote philosophers in their geographic works, the literature interests them more and more and there is a type of widening of the field and they act less as economists and are interested in everything that refers to space, all spatiality and especially the spatiality of the occurrence of the facts. I can give the example of coal and the capacity of vessels. Recently a study was done into what types of ships are used more often for a type of container... and there is not much to do in terms of concept. I can say that this is not the maritime geography of the 1970s, but the geography of the 1960s which translated this way and has not completely disappeared. This way of seeing things, still resists. It can be considered as a clear regression and bring into question the global gradation of more relational division of global space with a retrocession in certain areas. It is necessary to understand that the logic of this retrocession is above all economic, which influences the formation and indicates an evolution. I think it is hard to say where new geography is practiced. It can be said that there are some places remaining for modern geography. However there is not much external aggregation although there is no more internal aggregation, this is also linked to the nature of proofs and is important due to its consequences for the whole education system. On the other hand, there is still a conservative bunker very linked to a traditional cartographic approach, cartography for Africa and it has not assimilated the changes in cartography, it does not accept the idea of geography as a social science, in their small "world". But the number of bunkers of traditional geography, it can be said that there is a movement everywhere in universities, in laboratories, in unexpected places more than anywhere else. There are also those that play an important role in geography such as Hérodote journal, *Espace et Societe*, and *Espaces Temps* and there are also some remaining *fiefdoms* and others that appear like RECLUS and the people who work with Paul Claval. There is evolution, new things and ideas appear, but there are also addresses that are *static*, laboratories in a region where one can find all the specters of geography and also of space, although there are also interesting things. And there are places that are demarcated like the *Sciences Politiques* where I work. There is political geography and the director Alain Lancelot is a political scientist who loves geography and denies Siegfried's tradition. He was geographer but denied geography and took refuge in geopolitics, political imperatives. And so there is a renewal of Siegfried's tradition of the relationship between space and politics, and geography accepted this well and it develops and there are a good number of geographers who are teaching and producing, and there are new places emerging, etc.

What has changed globally today is that geography has a better image outside and the demand for the work of geographers and for the geographers themselves; this is an idea that is no longer strange in the world outside. Kilometers are crossed. It is not hard to find the political center; but when it comes to political geography it is something else. One thing is the Pakistani question, another issue is the desire of Serbia to build independent states;

another issue is the part of province, power is power. The space of Geopolitics and political space are not the same spaces. Thus the importance of distinguishing between the bases.

Following on for this I ask Jacques Levy to talk about the *Espaces Temps* journal, its importance for his formation, his life considering that he is one of the contributors to this journal. He answers:

When I became one of the contributors to *Espaces Temps*, it already had its defining character, a collective character. There is no boss at *Espaces Temps*, there is no employer. So democracy is tiring, in a way, not as a system but as a way of getting things done, although it is a precious advantage. I believe that if *Espaces Temps* were not that way, the results would be very different, because everyone is on a committee, no matter who they are. There is no cooption and there is diversity on the committee, members necessarily have different and often divergent views and in general decisions are taken by consensus; when there is divergence each person takes a long time to explain. And that creates a kind of natural continuous training, from everyone for everyone and personally, much of the little I know of other social sciences, apart from geography, I owe to *Espaces Temps*, because it is there that you have access to important books. When these books come out in various disciplines, we learn about them, although unfortunately in geography there are fewer new publications. This makes it easy to find out what important things are being produced and how things are done, what the practice is, and how it develops. But I know that if there were not people regularly talking about what is happening in sociology, in economics, in the Southeast, what is reproduced, etc. I would not know much. There is freedom to speak and there is no power game and no one has power over others. This does not happen very often in the universe of research institutions. There are often, unfortunately, authoritarian temptations and even in large projects there are tyrants *enlightened* by petty ambitions. This is the worst thing, there is everything to lose, the little things, the little signs of power and that, personally, does not interest me. So I have distinct relationships, I am distant from my research institution where teams are often formed where people choose among themselves and forget to work with someone. In *Espaces Temps* it is a right; people truly support each other because of the democratic sense. It is stressful because it is a handcrafted work, which has no automatic financial support and everyone has to do their part; those who are there, truly want to be there, so I dedicated my thesis to this group of people at *Espaces Temps* because it is the place where people have made an impression on me and have played an important role in my intellectual journey, and these people, *Espaces Temps* is my legitimate space, and in any case it is comfortable.

Enthused by his ready answers, I ask further about previous answers. I return to the fact that he knew the AGB (Association of Brazilian Geographers), "you also asked me about the actions of the AGB in Brazilian Geographical training with regard to the formation of students and exchanges in geography in Brazil. Developing a parallel what do you have to say about the Association of French Geographers or the French Association for the Development of geography?" He instantly replies:

The institutions of French geography are very complex and unfortunately there is not really an equivalent to the AGB. The students have overtaken the masters, be they P. George, Monbeig. The AGF has played a more dynamic role in the past, for sure. Today it is just a marginal institution. But it is surprising that they are the institutions which have a vocation to represent geography with the authorities, which are meant to represent the Geographical vision of the outside world. The UGI (International Geographical Union) and French governmental institutions, for example. And indeed there are two exceptions to this dichotomy, reflected in the division that exists and should not exist in French geography.

These two associations are the AFDG and the National Committee of Geography. The National Committee was the only one that existed in the past and it has been accused by some geographers of not meeting their expectations, not being democratic, because it is necessary to hold the title of *Doctoratd'Etat* to enter this National Committee. And that meant that you could not choose to enter, it was discretionary, it was totally undemocratic, and quite logically this National Committee had a very conservative ideology, notably in terms of power and internal discipline. In 1968 Full Professors were tyrannical to assistants.

I ask whether this would still be a colonial influence in geography. Levy confirms and continues:

Certainly. It was like the domination of the black slaves. And there was a kind of humiliation of university students and I was on the other side. In 1980/82 an organization was created, the AFDG (French Association for the Development of Geography), which was open to all geographers, covering and including secondary education and students. I was part of the founding group, I was the first vice-president of the AFDG in the first term, and we were inspired by the AGB. Upon my return to Brazil in 1982, I was particularly impressed by what I saw in Porto Alegre, and I said to myself "this is great, we should have something like this". This organization was the AFDG, with the idea of an open association that is secure, to promote all new ideas, all the innovations, all the titles of French associations for the development of geography. So this means that the National Committee did not contribute to the development of geography. Now there is a kind of duality, but it is not really a rupture, the relationship between the National Committee and the AFDG is peaceful. The AFDG holds an annual event called GEOFORUM and is gradually occupying, by the will of geographers, the Geographical Conferences that are the National Committee's usual events. Currently the Committee has been pacified. Younger scholars participate in it, and they now form the AFDG. They have important positions of power, they are professors, not only assistants, and masters of conferences. And so the French University has evolved in the sense of a lighter power, changes in the hierarchy and geography as well. The fact is that the ideological positions that were once very strong are weaker today, and especially in geography, ideological positions and theoretical positions have changed extensively. It is the result of the action of the political leftwing and also the progressive trend in at a theoretical level. Now things are simpler, more evident and positions between the right and left appear clearer and there are progressive and conservative theorists in all political camps. So it is not enough not just to be "leftwing", for example, to produce an innovative vision of geography.

As a former ex-president of the AGB, I tell Levy that I enjoyed understanding this issue, because the AGB also had an important role in my training. For me, the AFDG does really seem similar to the AGB. I have noticed that most of the Brazilians who go to France do not know about the association. Levy replies:

It is possible that the Brazilians do not know it. As for the register of entities there are not any books. But the other one as well, the National Committee is unknown, totally unknown outside and I ask myself how a small group, a limited group like RECLUS can be more well known outside than the AFDG. But even when it comes to the National Committee, the conservative view of geography, who ultimately are the actors when it comes to changing secondary school programs, there is always a particular conservative lobby and they are well organized, better structured than the Association of Secondary School Teachers which has another extremely powerful lobby, which generally opposes any change. They are more effective than AFDG and the National Committee, which may be all generalists but are not sufficiently mobilized. Another activity that drives geographers annually is the International Festival of Geography of Saint Die des Vosges and the current issue is organizing a symposium, like a festival that brings together all geographers, this is an occasion for a general public, anything which will work, which seems to be external to the heart of the discipline and resemble an unofficial manifestation of geographers aimed at university students. Anything that evolves, so that geographers perceive themselves as a *tribe*, as a body, as if from time to time they have a kind of corporate passion; but essentially, they are more interested in their little "areas" of work than in the abstract identity of the geographer, and geography, is happier and stronger than before, they no longer need ritual to assert themselves as geographers.

I finished the interview and pleased with the result obtained. I thank the availability and kindness of Jacques Levy.

It seems that the choice of Paul Claval and Jacques Levy fulfilled the goal of our research. They prove the impact that these two geographers have in the process of renewal of the relationship between France and Brazil. The tone of the interviews is evidence of the importance of these two geographers.

This selection does not intend to exclude other geographers who have developed activities in Brazil. On the contrary, we have cited Herve Thery, who is currently the director of the Maison de Geographie de Montpellier of the RECLUS group; given his role as a specialized researcher in Brazil, he is an obligatory figure when discussing the actions of French geographers in Brazil. In addition, the RECLUS group has a special agreement with IBGE, which reinforces the contact with Thery and his team. Martine Droulers, coordinator of CREDAL's Brazil Group at IHEAL, is another very experienced professional in Brazil; she has held an academic rank post at the Federal University of Paraíba and participated in several research teams with Brazilians in bilateral agreements. Lately, she has developed research with a team from Rio de Janeiro regarding technopoles and spatial organization and another on agricultural business. A disciple of Monbeig, she developed her doctoral thesis on agricultural expansion in Maranhão and maintains a consistent production about the country. In Paris, CREDAL's Brazil Group is an important support point for Brazilian researchers. Jacques Malesyeux's work with urbanists linked to the University of Brasilia is also noteworthy.

Pebaylle could be mentioned among those scholars known to have a history in the framework of relations between the two countries. However, we insist that in the case of the two geographers chosen, and whose complete interviews are included herein, we see different situations: Paul Claval is a recent inclusion in our geography, at a time when his reflections have reached maturity. He is a professional born in 1931. Although he declares that his emotional bonds and/or interest in Brazil began early in the period when he was at university, his supervision of theses occurs after 1980. His books are always cited by Brazilian geographers, making him a sort of obligatory reference in various themes dealt with by geography. His varied work includes issues related to geographical epistemology, the history of geography, the geography of power, urban geography, teaching geography, and cultural geography. Regardless of his political positions, his works are cited in Brazil by geographers from all political and party trends. His greatest expression in our geographical setting, as regards relations with France, stems from recent years, guaranteeing him an outstanding position among the advisors for Brazilian's theses and research.

Regarding Jacques Levy, the historical time frame is quite different. His politics and production guarantee him a comfortable position in the Brazilian geographical environment. We believe that a flow has emerged around Jacques Levy who has demonstrated a desire to strengthen his relationships with Brazilian geographers and assumes a posture of the utmost respect and admiration for our geographical production.

References

Berque A (1992) Espace, milieu, paysage, environement. In: Encyclopédie de Géographie. Economica Paris, p 358

Claval P (1987) A Nova Geografia. Livraria Almedina, Coimbra, 158 p

Chapter 6
Brazil in France

Abstract The chapter's approach reveals how Brazilian geography comes to France in the period being analyzed and emphasizes the invitation that was made to Lysia Maria Cavalcante Bernardes, to minister the course "Types of urban networks in Brazil," at IHEAL, Institute of High Studies of Latin America, in 1967. The title reveals the extent to which Brazilian geography was up to date at that time. The Brazilian guest worked on a theme which had a conceptual structure which was still regarded as a novelty in France. A rare moment of synchronicity between the two countries occurred, in terms of themes studied. In Toulouse, in this year, Milton Santos lectured on "Population and Food." It is important to emphasize the way Brazil was discovered by France, as is revealed in research carried out in the INTERGEO Bulletin Number 10, of 1968 in which Brazil was the subject of books, research, and films.

Keywords Discovery · Actualization · Research about Brazil · Professional exercise

6.1 Brazil in France

> I would say that compared to the interest in Brazil in the 1970s, the current interest is less. It seems to me that interest in Brazil today is part of a more general interest in Latin America. I think that until today, despite everything, Brazil's image in France is influenced by what General de Gaulle said long ago, "Brazil is not a serious country". So, both here and there, the 'impeachment' was actually the fact that changed this perception a little, and people thought "Oh, in one of these countries there is a place where they are capable of doing something politically coherent".

These are words of Marion Aubrée, an anthropologist from the Center for Research on Contemporary Brazil, at the Maison des Sciences de L 'Homme da Ecole de Hautes Etudes en Sciences Socialles. She is a professional with many years of experience in research in and about Brazil. She is also the coordinator of a CAPES/CaFE/CUB Agreement. In an interview conceded to the author in January

© The Author(s) 2016
J.B. da Silva, *French-Brazilian Geography*, SpringerBriefs
in Latin American Studies, DOI 10.1007/978-3-319-31023-7_6

1993, she synthesized the situation of the Brazilian presence in France and its effects, in addition to discussing the formation of Brazil's image in that country.

Jean Tricart, the renowned geographer from Strasbourg, known for his excellent research on the Amazon, said the following about Brazil:

> The Collor affair promptly, boldly, and honestly conducted raises some hope: the implementation of the 1988 Constitution is advanced compared with the legislation of other countries, but it will not have a place without a minimum of prosperity.

These points of views open the discussion surrounding the effects of the relationship maintained by the two countries, emphasizing the opinions given by researchers and other professionals whose activities have generated images of Brazil.

Brazil's image in France was constructed by scientific knowledge based on the arrival of travelers and artists. The contact with the tropical world, meeting with a diverse and adverse culture, and in some cases the known and established, provoked a big interest in Europe and France in particular.

France established itself as a cultural reference in Brazil, being hegemonic in certain periods. France with the charm of its language, the fame of its cuisine and the finesse of its salons was very successful in Brazil. The nation became a strong follower that offered many market opportunities for a country going through a period of industrial expansion and was facing strong competitors.

In geography, Brazil began a more frequent contact with French geography through Delgado de Carvalho who introduced a scientific approach to geographic thought and reflection.

In an interview given to the *O Estado de São Paulo* newspaper's *Caderno Cultura* on September 13, 1986, Herve Thery answers the question "What is the image of Brazil in Paris, especially at the Ecole Normale Superieure in Paris?"

> I cannot answer for the School, which also has not adopted Brazil as a compulsory subject. The presence of students is optional at the general education seminars about Brazil that I have given here for two years with the economist Alain Zantmann. However, it is clear that the interest in this initiative is increasing more and more. I do not know if a seminar on another country would arouse the same curiosity. In fact, between France and Brazil there is an old relationship of enthusiasm and mutual attraction that does not galvanize crowds on either side of the Atlantic, but they are sufficient to ensure the operation of exchange centers as inspired by the seminar by Ignacy Sachs at the School of Higher Studies in Social Sciences in Paris, or at our workshop at the Ecole Normale. Interestingly, it is noteworthy that the history of sympathy, of seduction between the two countries rests on a small number of misunderstandings, slightly false images entertained by both sides.

The content of the interview allows us to include it to explain what these misunderstandings are in the specific case of geography. For illustration purposes, we have selected some impressions and records that discuss relations between the two countries from the CNRS's INTERGEO Bulletin, which began publication in 1966.

The INTERGEO is organized by year and by universities and/or research centers. In the case of doubts, centers and universities in Paris and its surroundings were sought for so as to identify, as far as possible, the name of the student and confirm the nationality. Another case which demanded much acuity in the research and registering of information was that of French researchers and students who did their research in Brazil. The same criteria were used, and in this case, the nationality of researchers and French supervisors linked to the theme was sought. In the list presented, the selection was made according to the name of students and their theme so as to try and locate Brazilians, as there was no identification with regard to the origin of students.

The main subject of the INTERGEO No 1 (1966) was the Program for the Application of Reform in Higher Education prepared by a committee appointed and consisting of eight historians and eight geographers. The geographers were Derruau, Dresch, Juillard, Papy, Pinchemel (absent), Roncayolo, Taillefer, and Veyret.

In this post-1964 period Brazil is not mentioned, no records were found.

The second INTERGEO volume (1966) was characterized by a globalized aspect in terms of the tropical world and underdevelopment.

Of the teachers with interests linked to Brazil, working on tropical issues in the academic year 65/66, there was Guy Lasserre in Bordeaux, who taught tropical geography and tropical Agriculture, among other disciplines. In Caen, M. Andre Jornnaux taught a course on the geography of tropical America. These courses were actually offered by several French universities during this period (Dijon, Lyon, Nancy, Paris, Strasbourg, Rouen).

In the same year in Montpellier, M. Deffontaines organized a course on Latin America and in Paris M. Monbeig taught geography of population and agriculture in tropical countries.

While most professors were focusing on regionalized analyses, in Paris P. George discussed urbanization issues and M.M. Rochefort worked on study methods of the urban framework, urban population, and urban networks and gave another course on the production of the goods of equipment in the world.

In Toulouse, a Brazilian, Milton Santos, gave a course on the "geography of Population—Underdeveloped Countries."

The political repression was imposed by the military coup of 1964, if on the one hand it caused the loss of thinkers of the quality of Milton Santos, Josuéde Castro, Celso Furtado, and many others and on the other hand it allowed the world, through France to perceive the intellectual capacity of Brazilians. According to Ignacy Sachs, coordinator of the Research Center on Contemporary Brazil, in an interview given to the author in January 1993:

It was Milton who contributed the most towards the knowledge of Brazil here, partly because of his many years of experience in the French University.

Marion Aubrée adds:

Milton Santos was in exile here for a long time, and it was not just through his science that he enabled the French to get to know the reality of Brazil better. I think it was also his person, his subjectivity. He was living in another country; it was an exchange, a mutual enriching, not only through science, but also through the customs of each one, that is an enrichment of knowing Brazil's culture better. Getting to know the other culture's reality and what is eternal in that culture.

French academic production in the field of geography did not privilege Latin America. One of the few works dedicated to the Latin American continent was by M. Dollfuss, *The Central Andes in Peru and their Foothills*, a thesis defended on February 12, 1966. In the same year, Helene Riviere D'Arc, who would become well known in Brazil, prepared her *Diplome D'Etudes Superieur* entitled *Study of a Mexican Suburb*, directed by M. Monbeig.

Still in the field of geographical research, in 1966 Dollfuss and Rochefort worked on a project about *Medium-sized Cities and their Regional Action in Latin America*.

A treatment of an exclusive theme on Brazil appeared for the first time in the INTERGEO in Poitiers, with the third cycle theses of F. Kott, entitled *Brasilia and its Region* supervised by M. Robert and defended in October 1966.

An interesting registry in INTERGEO is in item five, also published in volume two of 1966 about *missions and movements*, regarding visits by French and foreign colleagues. In that academic year (September 1965–September 1966), M. Guy Lassere travelled between October 26 and December 15 of 1965 to São Paulo to teach at the Institute and Department of Geography at USP.

In the same year, Bordeaux received a visit from Milton Santos from the University of Bahia who was in Toulouse at that time. Regarding M. Santos, Lyon refers to a conference on "The original characteristics of agriculture in the Brazilian Northeast," to be held in 1966.

In Reims, Demangeot planned a journey to Mato Grosso sponsored by the FAO.

Another very interesting registry for the geographical community that permitted the location and tracking of events was section IV that contained a calendar of national and international events, conferences, and colloquiums. It recorded the International Conference of the CNRS, between the 11 and 14 of July, on the theme "The Agrarian Problems of Latin America."

Also of significant interest to the Brazilian geographic community was the tracking of activities that have Brazil as the center of interest. In its first numbers, the INTERGEO newsletter records a timid presence of Brazil. Themes relevant to Latin America were addressed as a block. What happened is treated in a general way. There is no consideration of the singularities and differences of Brazil in the Latin American context. The absence of Brazilians recorded in France developing research activities linked their country's needs, is due, we believe, to the political situation that the country was going through at that time and that would last for almost 25 years. Thus, vols. 3 and 4 do not include any subjects of interest to our area of research.

In the records for the year of 1967, Bulletin No 6 contains a theme that is worth registering. The thematic analysis was centered on the Latin American context. However, the title the "Tropical World" persisted. Among the teachers, Guy Lasserre was teaching "geography of population and food geography in tropical countries" in Bordeaux. Deffontaines taught a specific course on Latin America, in Montpellier, entitled "South America."

After this edition, there was a change in the treatment of geographic themes according to the following theme: In IHEAL, M. Rochefort, focused on "comparative case studies of *d'amenagement* of territory" and "the great problems of human geography in Latin America." In the Rue Saint Jacques Geography Institute, the novelty was the "Study of Cities," in the form of a course taught by Pierre George. Also under the responsibility of M. Rochefort at the same Institute, there are records of new concepts being studied in the courses "Methods of Urban Networks Studies" and "Urban Functions." In Saint Claude Normal Superior School, Pierre M. George taught a course on "The Popular Democracies," a theme which was addressed in a significant number of works.

Brazilian geography reached France during this period through Lysia Maria Cavalcante Bernardes, with the course "Types of urban networks in Brazil," given at the IHEAL. The title indicated the level of actuality in Brazilian geography at that time. The Brazilian guest teacher worked on a theme that had a conceptual structure new to France in that period. There was a moment of rare synchronism between the two countries in terms of the themes being dealt with.

During this year, in Toulouse, Milton Santos was lecturing on "Population and Food."

Discussions based on the theme of *l'aménagement*, developed by B. Kayser, gained weight in the 1970s along with the concept of the organization of space. Berque[1] refers to the use of the concept of space by Jean Labasse in his book *The Organization of Space*, of 1966.

> This title is evidence of the effect of the appropriation characterized by a scientific discipline of a term formerly thought to be effectively *banal*.

Among the publications, Rochefort and Monbeig's book *The America of the South Atlantic published* by Magellan is worth highlighting.

Advances in the conceptual structure of urban geography are evident in the book *L' Armature Urbaine Française* edited by PUF and written by Rochefort and Hautreux.

At this time, F. Kolt remains the only researcher on a theme exclusively about Brazil, with his thesis "Brasília and its region." The new capital city became a center of interest for geographical research.

Rochefort travelled to Brazil as a member of a technical cooperation mission, during July and August of 1967, as part of the exchange program.

France rediscovers Brazil! In INTERGEO No 10 of 1968, Brazil which had been the theme of books, research, and films during a long time had been kind of forgotten.

---

[1]Berque (1992, p. 358).

Previously, in 1900, Elisee Reclus wrote *The United States of Brazil* which was published in Rio de Janeiro by Garnier. In 1909, Pierre Denis made Brazil the subject of his book.

After the founding of USP, the University of the Federal District and IBGE, the country achieved considerable visibility. The presence of renowned geographers such as P. Deffontaines, Monbeig, and other teachers who would go on to become the most famous intellectuals in France had repercussions for the construction of Brazil's image overseas. In the field of geography, the arrival of the two pioneers was the beginning of Brazil's intense relationship with France. INTERGEO only began in 1966. Until that year, given the lack of systematic records, the only possible way to analyze the presence of professional geographers from France who visited Brazil is via a chronological organization of their stays and permanence, mapping their areas of work and researching their production.

From 1966, as a systematic bulletin of the activities developed by geographers in France, INTERGEO provides meticulous elements that allow more refined interpretations and analyzes according to the interest in the subject and the researcher.

In affirming that France rediscovered Brazil, we are basing our analysis on two factors. First is the scarcity of researchers in the 1950s, and second, the almost complete absence of Brazil as a theme of courses or research in French universities.

An analysis of the earlier numbers reveals this absence. On the other hand, after 1968 (the academic year of September 67–September 1968), Brazil emerges as a theme in almost all the centers studying geography in France. There is also a considerable number of Brazilians developing their research in that country. In Bordeaux, Milton Santos lectured on "geography and urban economics in developing countries Brazil." Together with B. Kayseer, now at the Institutd'Etude du Developpement Economiqueet Social—IEDES, he gave the course *Urbanisation et Organization de l'Espacedans les pays envoie de developpement.* In the same year, still at IEDES, another renowned Brazilian Josué de Castro gave the course *Quelques problemes alimentaires dans le monde contemporain.*

Interest in Brazil grew during this period. The country is in evidence; it is fashionable "*c'est la nouvelle vague.*" The best way to perceive the various approaches is to directly or indirectly organize the INTERGEO list by universities, centers, or institutes. The following are the selected centers of research with the respective professionals responsible for the disciplines and their titles.

- **Clermont Ferrant**: Derruau worked on the theme "Japan… Brazil, British Isles."
- **Nancy**: Planhol lectured on *Le Bresil: l'habitationrurale.*
- **Nantes**: Mlle. Mesnard gave a course on *La France, Le Bresil, la Venesuela* and *La France, le Benelux et le Bresil.*
- **Orleans-Tours**: In the discipline of Agrarian geography, P. Fenelon worked on the topic *L'Afrique, Le Bresil e Le Loire.*
- **Paris**: In the geography Institute of the University of Paris, Monbeig taught *Le Bresil.* Rochefort continued his research on *Les fonctions Urbaines* and *Geographie des Activites Tertiaires.* P. George worked with the discipline of

Urban Studies. At the IEDES, the theme of underdevelopment appeared in P. George's courses on *Caracteres geographiques des pays envoie de developpement* and B. Kayser's "The geographical situations of underdevelopment." At the IHEAL, confirming the interest in Brazil, at that institute Demangeot lectured on *Exercices de Photo-interpretation sur le Bresil.* Rochefort developed the courses: "Human and Economic geography of Latin America" and "The Role of Cities in the Regional Development of Brazil."

- **Poitiers**, J. Cabot teaches *Bresil, Agriculture Tropicale.*
- **Strasbourg**, Gallais lectures a course on *Le Bresil Tropical.*
- **Toulouse**, Demangeot teaches *Geographie Zonale—Le Bresil.*

The change in position by the centers linked to teaching and research is clear with regard to Brazil as a center of interest, as manifest by the titles of various disciplines. It is evident that there was a political environment capable of showing that all this effort in France to make Brazil a center of attention comes at a time when domestic issues became more serious. In France, *May 68*—the student rebellion demanded a more open and active university. Here, students organized and mobilized against the repression of the military regime. The *March of the One Hundred Thousand* demanded an end to the dictatorship that had ruled the country from April 1964. The year 1968 was marked by the AI 5 (Institutional Act Number Five), and this was a time when a significant number of Brazilians were exiled or forced to live in hiding. In addition to offering political asylum to several fugitives, France was sensitive to the issue of the struggle to redemocratize the country, condemning the regime. Contradictorily, at least at first glance, during this period of strong political repression, Brazil experienced one of its most fertile cultural and intellectual moments. *Bossa Nova, Tropicalismo, Cinema Novo,* and large popular music festivals revealed a country searching for the new. The pulse of modernity inspired a strong debate in the country, integrating Brazilians who were prevented from using their communicative capacities freely and encouraged them to try new ways to circumvent the dictatorial rules.

The voice of the people is not silenced. The changes did not prevent our cry from echoing in other quarters. France, among other countries sympathetic to the democratic cause, was at that time one of our possibilities for condemnation and expression.

With regard to the INTERGEO Bulletin, the research indicated a greater number of courses and activities. In Poitiers, Kott finished his third cycle thesis entitled *Brasília e sarégion: Etude de Geografie Vilaine.* Completing the picture of this time, there were three pieces of research developed at the *Maitrise* level, the first by Ana Maria Montenegro, *Maitre de Conferances* at the University of Paris XII, Paris Val-de-Marne, "Regional Variations in Education Levels in Latin America and in Brazil in particular," the second by the French scholar Foucher about *Le Developpement du Reseau Routier au Bresil,* and the third by M. Ribeiro on *La ville et la Region du Bahia: Colonisation et Contatcts de Civilisation.*

As to publications, Gallais published *L'Amenagement Agraire de la Serra de Baturite—Bresil*with Cahiers des Ameriques Latines in 1968.

Regarding Missions and Journeys, Lasserre, from Bordeaux, went on a teaching mission to Brazil. The University of Lyon received a visit from Milton Santos of the University of Salvador. The University of Paris informed of the visit of P. George to Brazil, to give courses at USP in the months of April and May 1968. Toulouse registered a mission to Brazil in the summer of 1969. The following table contains the number of theses by the French about Brazil (according to INTERGEO).

Theses about Brazil by French students

Years	1966	1967	1968	1969	1970	1971	1972	1973	1974	1975	1976	1977	1978
Enrolled	1	1		6	1	1	3	2	1	3	4		3
Defended		2		2	1		2	2	1		2		2
Years	1979	1980	1981	1982	1983	1984	1985	1986	1987	1988	1989	1990	1991
Enrolled		2			1				3	3	3	1	
Defended		4						1	2	1			

Theses about Brazil by French students by advisers

Rochefort	06
Monbeig	04
Robert, Celso Furtado, Tricart, Leloup, Gaignard, Barbier	
Revel-Moroz, Héléne Lamicq, Di Meo	01
No indication	03

Theses about Brazil by French students by universities

Paris (until 1969)	08
After 1969	
Paris I	08
Paris III	05
E.H.E.S.S	02
Paris XII	01
E.P.H.E	01
Total Paris	25
Interior	
Strasbourg	03
Toulouse	02
Rouen	02
Poitiers	01
Reims	01
Lyon	01
Aix-em-Marseille	01
Pau	01
Nantes	01
Bordeaux	01
Total	14

Themes studied

Urban geography, city and region, tertiary	14
Agrarian geography, expansion of agricultural frontiers	
Pioneering front	08
Industrialization, "amênagement"	07
Transport	03
Cerrado, Caatinga	02
Other themes	04

Approach/location

General	
Brazil	08
Regional	
Northeast	04
Amazonian	03
Center-west	02
South	02
State	
Minas Gerais	02
Maranhão	02
Paraíba, Bahia, São Paulo, Rio Grande Do Sul, Rondônia	01
Valleys	
São Francisco	01
Cities	
Brasília	02
Recife	02
São paulo	02
Salvador	02
Rio de Janeiro	01

Source Bulletin INTERGEO, Paris, CNRS, 1966/1991, vols. 1–104

Reference

Berque A (1992) Espace, milieu, paysage, environement. In: Encyclopédie de Géographie. Economica.. Economica, Paris, p 358

Chapter 7
The Production of Theses by Brazilians in France

Abstract On the political stage, scenes of accelerated modernization of the country were being drawn, based on programs of bilateral cooperation. This was a strong call, and in this call, France stood out. The situation of dependency remains, and France calmly takes advantage of its status as a role model for the world in diverse sectors. There is evidence of French supremacy at various levels in the relationships established in the fields of formation or internship in Geography above all those maintained between teachers and students, researchers and interns, and so forth. This becomes more intense in the period following 1934 when our sense of lack was complete and the discourse concerning progress takes it shape. The need to form personnel able to structure and maintain a satisfactory academic environment in Brazil involves nearly all areas of knowledge.

Keywords Modernization of Brazil · Standing out · Model · Progress

Naturally, France became a magnet for Brazilians working on research and theses as it had specialized centers and qualified faculty and it therefore provided the conditions that led to and perpetuated the academic ties established institutionally via covenants and technical–scientific cooperation agreements between the two countries.

In the political context, scenarios of accelerated modernization were designed for the country that were made viable through bilateral cooperation programs. The call was very strong, and in this call, France stood out. Contemplating this subject, Velho said:

> ...it can be said that the Brazilian elite has always been fascinated by 'modernization'. At least in the sense of having countries considered more advanced as a model, above all (in variable order) the United States, France, Great Britain and Germany. In politics, some of the principle episodes (such as the Proclamation of the Republic in 1889 and the revolution of 1930) can be read as changes in the hegemonic models. Although this does not exclude coexistence, France, for example, maintained more or less the same privileged position with reference to *culture* and customs.[1]

---

[1]Velho (1992, p 199).

© The Author(s) 2016
J.B. da Silva, *French-Brazilian Geography*, SpringerBriefs
in Latin American Studies, DOI 10.1007/978-3-319-31023-7_7

The situation of dependence persists, and France calmly enjoys its position of a model to the world in several sectors. In the relationships established in training or internships in geography, French supremacy is evident at various levels, above all, in those between professor/student, researcher/intern, and so on. This intensifies after 1934, when our scenario of neediness was complete and the discourse on progress took shape. In this regard, comments and statements by Orlando Valverde lead us to think that closer academic ties would be more interesting at the time that we had the conditions for an exchange based on our reflection and production.

> Despite France awarding scholarships and research internships to Brazilian student geographers, it is undeniable that the close ties established before the Second World War, have not been sustained until the present day. However, it would be easier now as higher education in geography has produced results in Brazil[2]

This situation of dependence, the incapacity to reflect on geographic theory and especially on our spatial reality, will be overcome later. The formation of a competent framework of Brazilian geographers has created the conditions essential to contemplate the formation of a Brazilian geographic school. The questioning about the training capable of structuring and maintaining our academic environment in a satisfactory way involves practically all areas of knowledge. When questioning our intellectual universe, our centers of excellence in scientific production, and our ability to think and reflect about the world and in that world, Brazil, we become aware of our shortcomings, but we also realize the significant advances we have made. In this respect, Micelli affirms:

> For a long time Brazilian intellectuals were reduced by the quality of the scientific ideas transmitted by renowned French researchers like Fernand Braudel, Claude Levi-Strauss, Roger Bastide, Maurice Bye or François Perroux... today, Brazilian sociologists, economists or historians do not have anything to desire from their French counterparts.[3]

Brazilian geographers could easily have been cited by Micelli. For some time, we have had a qualified body of professionals with a production that often extrapolates the limits of our borders. At the same time, cooperation and exchange programs with their asymmetrical design meet with resistance, create resentment, and occasionally provoke nationalistic policies and measures with a strong conservative tone.

We hope for a cooperation program that overcomes the practices of supremacy and domination by one party and avoids fierce competition between the partners. In this senses, Berque points out that:

> Cooperation does not consist only of a transfer neither of objects or practices, not even methods but of exploratory procedures postulating a dialogue between civilizations. This dialogue does not mean that all the elements are present in our own system. In the medium and long term and from then on, the creativity of French thinking and research and even

---

[2]Valverde, Orlando. In: op. cit. (T.A) p. 8.4.

[3]Micelli, Sergio. In: op. cit. (T.A) p. 265.

French society in all fields will supply the material that will raise the demand for cooperation.[4]

In this respect, there is strong evidence that things are moving in that direction. The professors and researchers interviewed, for the most part, consider Brazilian geography mature and in a more comfortable situation in the relationship of exchange.

Critical thinking presupposes seeking unprecedented alternatives to improve the quality of Brazilian geography. The emergence of a new creative professional committed to an original and critical geography is only possible with high-quality graduate and postgraduate programs. In Brazil, the expansion of postgraduate centers of geography has been growing considerably. The creation of the Postgraduate Geography Association will certainly make research that systematizes and disseminates the integrated information of these postgraduate programs viable. The agencies' funding research and the training of senior staff have encouraged *sandwich* grants, increasing the completion of postgraduate courses in the country, offering nonetheless conditions for the postgraduate student to study part of their course abroad. This modality is becoming more common, without hindering the approval of grants for complete postgraduate studies abroad.

This discussion comes to the fore at this time, when we are discussing Brazilians' geographical production in France by evaluating the historical process that generated the flow of Brazilians to that country. This flow was consolidated due to our shortfalls and was added to by professionals from other fields, particularly architects and urban planners in France who were drawn to do their theses in geography. The prestige of this science in France elected leading figures such as Monbeig, Tricart, and Rochefort, who also stood out for the high number of theses they directed of professionals from other fields of knowledge.

In an interview carried out in Paris in January 1993, Milton Santos provides excellent information in a rich testimony as in this passage referring to the various phases of relations France/Brazil:

> In this phase there is the multiple influence of geographers from various parts of the world including the French, because of the seductiveness of French culture, the ability with the French language of those generations, the French government's interest in 'pushing' French culture; all of this, and above all the strength of French geography, because the French never presented themselves to foreigners as divided, they appeared as a school, despite the usual disagreements. So in this case you would have to see who came, how many grant students there were in France, there were much more than in the phase of infancy and less than in the phase of maturity. There was a quantitative increase in the Brazilian presence in France and the French presence in Brazil. There was an aggressiveness on the part of provincial universities, starting with Strasbourg, where Tricart was and where Rochefort, Monbeig, Dollfuss passed through along with Juillard and Sauter; there was a strong contingent of specialists in the tropical world. Toulouse was emerging, Bordeaux sought to maintain its old colonial and tropical vocation and then Clermont-Ferrand associated itself with these universities. For example, there was an opening in the range of personal influences and many more geographers came here (to France). In Brazil there was also an

---

[4]Berque J (1982, p. 22).

expansion of university life and later on the institutionalization (before institutionalizing) of career training, that is there are way that of entering the discipline, the university that become institutionalized long before Caiena became institutionalized. This led to a dispersion that was reflected in the production, and only what came in the case of the youth, there was a unitary manifestation, first through quantitative geography, which is a unitary form, and then through Marxist geographies that also have a unitary form for both are totalizing, with a tendency to be totalitarian, sometimes demanding debate, sometimes submission, but they will play a very big role in the recreation of Brazilian geography so it can question itself as a discipline.

It is possible that the questions that Brazilian geography asks itself are connected to the affirmation of other social disciplines and the social hierarchy of disciplines and planning, how it began to be done. Economic growth must have had a role here too because when disciplines begin to enter the market then there is an exchange because they have to be reviewed to have a good price, to be more or less appreciated. I think that is what leads to this theoretical debate within geography.What does this have to do with French geography?

Then you have a search by those who had something to offer on a conceptual level. So concepts become central and the producers of concepts become sought after. And then there are extensions. It is in this sense that a Rochefort, a Kayser… all on Master George's treadmill, they are the link between two moments. They appear as the producers of concepts, a studied region is not in itself the study of a region, but is the region. The growth of geographers is confirmed with Peroux and Boudeville who make the link between geography and economics, although another line will also appear in the 1960 s and above all in the 1970 s, when all the great countries became concerned with technical assistance and international trade. They begin to have a larger role in intellectual cooperation and that was also the time of the increase of Brazilianists in all the countries, that is to say… at the same time that Brazil could not be studied by Brazilians because the regime did not like Brazilian social scientists, they preferred to send them abroad, there was a certain permission for foreign social scientists to come to Brazil. So that period, immediately preceding the establishing of maturity, was also a time when there was a double demand for French geographers. One from the geographers that was the demand for those who think up concepts and the other that came from France itself, which is the sending of Brazilianists, some of whom will have a determinant influence in producing maturity. Maturity is the phase in which Brazilian geography produces its own theoretical and methodological textbooks, a geography that discusses itself and that can be read in Portuguese, which is a very important phenomenon, because geography encounters a publishing market which is concerned with ideas that will result in a series of phenomena in international relations that have yet to be analyzed.

The brilliance with which Milton Santos gives his opinions and points of view offers us multiple elements to facilitate the analysis of Brazilian production in France. The way he crosses the structuring components of our geographical formation, indicating key people, retrieving the historical process and inserting the whole dynamic of our geography, touching on the influence of different approaches in Brazil's troubled political context that, is associated to a perverse process which by excluding various Brazilians, favors foreigners in the analytical readings on the country.

We were unable to locate a systemic source that would allow us to quantify the total and location of theses produced before 1966, the year in which the CNRS's

INTERGEO Bulletin began, which is why the option was made to structure the history of that period on information collected in interviews and others found in sparse works. The partial dynamic of the process of Brazilian geographical training and the search for references about a privileged historical moment was given to us by Pedro Pinchas Geiger, in an interview carried out in Rio de Janeiro in October 1993. Questioned about the significance of French geography in his professional life, he said the following:

I sat the university entrance examination in 1939 and the course would start in 1940, the year of the Second World War. The Faculty of Philosophy had been created a few years earlier, that is, I was in one of the first groups of a formal university course called geography, when geography was allied to history. When it was created, actually until the Second World War, French cultural influence prevailed in Brazil. French was taught much more in secondary schools than English very few people spoke English. The second language after Portuguese was French. We have French traditions. So, when the Faculty was created they used French professors for several subjects. It was not just in geography. Antoine Bon was my professor for ancient history, Tapier taught modern history; Jacques Lambert of the *Os Dois Brasis* taught on the sociology course, there was another for mathematics, for philosophy. In other words, French influence at the time prevailed in the National Faculty of Philosophy at the University of Brazil, in those days Rio de Janeiro was the capital of Brazil. Deffontaines and de Martonne's visits to Rio were critical as they influenced the founding of the IBGE.

The creation of the National geography Council comes from a request that Brazil join the UGI. It is clear that the true meaning comes after. The creation of the IBGE was another thing, but this historical episode has to be registered. What happens is that France fell in 1940 and so the French professors who had come for a short stay, as was the case of Rio de Janeiro, I do not know, it may be the same for Monbeig, they were stopped from going back to France; and so they stayed in Brazil. In the case of Rio de Janeiro, Ruelan played a leading role and indeed we need to understand the following: the Faculties of Philosophy at that time were not research centers. The National School of Philosophy was created during the Capanema Reform to prepare secondary teachers, to give another level to secondary education. Only recently in geography, at USP, perhaps before, but in Rio de Janeiro the University has only recently become a place producing knowledge, in the sense of reproducing knowledge at the University, to have Doctors, Masters; for a while the production of geographic knowledge was focused on the IBGE, Ruelan understood this, and began to place his preferred studentsin the IBGE. The second group to join was Ruelan's, they were younger and went into this career without having any profession. Many of the first group went to the United States, because after the war there was an American entry, but those in the Faculty of Philosophy who had been influenced by the French professors, in the case of Rio de Janeiro, by Ruelan, were scholarship students in France. As I said, Ruelan was a geomorphologist. I myself was a geomorphologist early in my career and it is de Martonne's line that prevails.

Well, I had already had some involvement, as a student, in political movements in France, when I was a student there; I had access to the literature, to political economics and Marxism. I knew Lefebvre. My first contact with Lefebvre was in 1947. Not as a social scientist, He was a man for formal logic, dialectics. The effervescence in France, the effervescence of the left was very big, isn't that true? I already had a predisposition in terms of social vision, and no longer easily accepted the accepted vision of Vidal, and I already had a reaction, and this was reinforced by my stay in France. Well, when I came back, and as I told you, from this first wave of scholarship holders Helvio went to Strasbourg and Miguel to Paris in 1946 and Tricart was still studying, it appears that he was a professor in Strasbourg, but still had ties in Paris to do his doctorate. He had contact with Helvio and Miguel, who became his colleague and they formed a personal relationship, which explains what happened in 1956.[5] Elza went to Montpellier, I was sent to Grenoble to study Geomorphology... and Myriam went to Lyon.[6]

Geiger's testimony allows a partial reconstitution of the period 1940/1950 and enables us to understand and analyze the dynamics of the process of professional formation in geography in Brazil.

The interviews were a fundamental source in reconstructing this period. It was not possible to find systematized material which permitted the location, the quantification, and other information concerning those Brazilians who went to France before 1966, when the INTERGEO system was implemented by the CNRS.

The material collected offers several possible intersections and analyses. From 1966 until 1991, the inventory computes a total of 183 recorded in postgraduate programs in France. The total number of theses does not correspond exactly to the values in the table. There are cases of theses that had their name altered during the production phase, but it was not possible to check each case. In some situations, the student's name was not registered, and in others, the thesis director's name was missing, or the year of the defense, and so on. Even so, the source used after 1966 made it possible to get close to reality and an attempt was made to recreate the picture of the situation of Brazilian students in France.

Aiming to transmit the idea of the content of the Bulletin, some years were selected and an analysis was carried out to show the quality and quantity of the information available. The names of the people, the thesis advisers, and research titles were recorded in the same form as they were found in the INTERGEO.

Regarding Brazilian researchers in graduate programs, data for 1967 record the research carried out by Catherine Vergolino Dias, on "Agricultural Regions of the

---

[5] Author's Note The full names of the students who went to France after WWII: Héldio Xavier Lenz César, Miguel Alves de Lima, Elza Coelho de Souza Keller and Myriam Mesquita, Professor Sternbergs full name was Hilgard O'Reilly Sternberg.

[6] Author's Note: Prof. Milton Santos helped to clarify what the famous episode involving Professor Tricart in 1956, year of the UGI might be: "1956 was a very important year for international geography as it marked the shift in the balance of world geography, from Europe to the United States, it was the moment in which American influence in the UGI increases and one of the artifices of the moment was precisely Prof. Sternberg, who shortly after was named a professor in the United States. Sternberg organized a geography Congress in Rio de Janeiro, with some Brazilian colleagues, supported by the CNG (National Geography Council) and simultaneously he organized a Course in High Geographical Studies, which was a great event. In the organization of this course he carried out a type of ideological witch-hunt, he clearly vetoed the participation of professors with progressive thinking, except Prof. Monbeig, of course. I don't even know if Monbeig was giving this course, I don't believe so. So, his guests were several important people but Tricart for example wasn't invited by him, but he was invited by Miguel Alves de Lima, who was not progressive and invited Prof. Tricart to give a course alone at the old Lafayette. I have another example of Sternberg's ideological preoccupation, because when I had invited Tricart to Bahia, I was informed by my friends in the IBGE that Sternberg had obtained a memorandum from the Brazilian authorities requesting that Tricart should not be made welcome anywhere; but this memorandum was not obeyed by the federal workers in Bahia, with whom we had close relations" (Interview in January 1993, in Paris).

Amazon," and M. da Silva, on "A Socio-economic Study of a District in Salvador," both under the direction of M. Gallais, from Strasbourg.

In Toulouse, three Brazilians developed research for their third cycle doctoral theses: D. Lina de Brito, "Rubber in the State of Bahia"; P. Motti, "The Urban Network in the State of Bahia"; and S. Silva's thesis, "Regional Organization in the South of Bahia."

Milton Santos' presence as a teacher in Toulouse and his time in Strasbourg, where he did his doctoral thesis, helps explain the presence of four researchers from Bahia, living and studying in France in the postcoup period in Brazil.

In Paris, Maria Luisa Marcilio wrote her thesis *La ville de S. Paulo: peuplement et population de 1750 a 1850 (d'apres les registresparoisiaux et les recensementsanciens)*. Strasbourg continued to be the center with the largest number of Brazilians, at that time. Other researchers were as follows:

- Maria A. da Silva, *Les transformations du Reconcavo da Bahia sous l'influence du pétrole-Bresil*—Director: Gallais;
- Maria Novaes Pinto, *La culture du sisal dans l'est de Bahia-Bresil*—Director: Juillard;
- C. Peixoto, *Geomorphologie des environs de Salvador-Bahia, Bresil*. Director: Tricart;
- T. Prost Ribeiro Da Costa, *Aspects geomorphologiques du bassin du Mamanguape-ParaibaBresil* Director: Tricart;
- Catarina Vergolino Dias, *L'Agriculture Amazonienne* Director: Gallais.

During this period, two theses by Brazilians about Bahia were completed: the first by D. Lima Brito entitled *Le catchoudansl'Etat de Bahia*, in Toulouse, and the second by Sílvio Bandeira De Melo on *Le DécoupageRégional á Bahia*.

In this initial phase, there is a preference for northeastern themes with the supremacy of Bahia in the topics researched.

A general survey indicates that of the 183 theses registered, 80 were defended during this period. It is believed that the number could be higher. As to frequency, there is a no clear period in which theses are concentrated, although there is the suggestion of a higher volume of theses in the period 1984–1989, with a higher volume of thesis enrolled in this work period. The distribution in the period discussed is as follows:

Theses of Brazilians in France by year

Year	66	67	68	69	70	71	72	73	74	75	76	77	78
Enrolled		7	8	7	1	4	1/2	4	2	9	6	1	1/4
Defended				1			6	3	2	3	3		4
Year	79	80	81	82	83	84	85	86	87	88	89	90	91
Enrolled	6	2	1/0	6	6	9	1/4	8	1/7	1/0	1/2	8	
Defended	2	1	8	6	4	9	1/1	7	3	3	2	2	

Source Direct Research, 1992

Paris was the preferred center for Brazilians, with a total of 103 theses in the period 1966–1991. Until 1969, the record generally indicates Paris, since until this date the University was unified. The table is as follows:

Paris	
Until 1969	09 students
Until 1969	94 students

Distribution by university	
Paris I	
Paris XII	38
Paris III	21
Paris IV	16
Paris VII	07
Paris X	06
Paris	02
Vincennes	02
E.P.H.E	01
Paris VIII	01

The first three universities had 75 matriculated students, showing the characteristics of centers with specialized nuclei. The University of Paris I brings together a significant number of specialized professors, including some with a large number of supervisees like Michel Rochefort. In this period, there was also the late Pierre Monbeig, who passed away in 1987, and he does not appear on the list of doctoral supervisors, but he was in fact one of the mentors sponsoring the bilateral programs involving the two countries. The University of Paris III also counted on the presence of these two teachers as part of the staff supervising Brazilian students.

Interior	
Toulouse	17
Strasbourg	12
Bordeaux	10
Montpellier	05
Rouen	05
Grenoble	03
Poitiers	02
Aix-em-Provence	01
Brest	01
Caen	01
Lyon Nancy	01

Source Direct Research, 1992

The tables reveal a distribution that is consistent with the general opinion one has about the destinations of Brazilian's coming to France. In some of these centers, there are geographers working who possess and maintain affinity levels making them attractive to foreign students. Some of these professionals become so well known in Brazil that the center is often confused with the professional, a kind of concreteness that personifies the place. This is what happened with Tricart, who for most Brazilians is synonymous with Strasbourg, like Lasserre with Bordeaux and Kayser with Toulouse. This identity permits the construction of links with these centers, with which there is some continuity.

Theses of Brazilians in France by advisor	
Professor's name	N° of theses supervised
Rochefort	27
HeléneLamicq	09
Jean Gallais	07
Paul Claval	06
C. Colin Delavaud	06
Bernard Kayser	05
Guy Lasserre	04
A. Colin Delavaud	04
Jean Tricart	03
J. Malezyex	03
Bataillon	03
Ziv	03
J. Hubschman	03
Babanaux	03
Le Coz	03
Juillard	03
Noi, Bonnefont, Prats, Bonnamour,	
Hélène Riviére d' Arc, Chonchol, M. Durant	
Dastes, Dupuy, Sachs, BeujeuGernier, Coing	
Lacroix	02
Galabert, Nonn, P. Michel, Raymond,	
F. Mauro, Koechin, Revel Moroz, Prud' Homme	
Sternberg, Burgel, Rougerie, Fournie, Chaline,	
Noin, Cabannes, Saussol, Pébayle, Lacoste,	
Baurricaud, Huetz Le Lemps, Joly, Peirre	
George, Pinvhemel, Leloup, Jan Dessau	01

This list covers practically all the professionals who came to the country to do research and highlights the importance of M. Rochefort who declared in his interview that he had travelled 26 times to Brazil and supervised 27 students. This search confirms the affinity, competence, and recognition of his work.

Theses of Brazilians in France according to approach/location	
General	
Brazil	31
Developed, subdeveloped, and developing countries	04
Regions/large spaces	
Amazonia	03
Northeast	03
South	02
Southeast	01
Recôncavo baiano	03
São francisco Valley	01
Parnaíba Valley	01
Paraíba Valley	01
Mamanguape Valley	01
Carajás project	01
Belém-pará highway	01
States	
	15
Bahia	05
Minas gerais	04
Paraíba	03
Ceará	03
São paulo	03
Acre	02
Rio de janeiro	02
Santa catarina	02
Piauí	02
Goiás	02
Rio grande do sul	02
Espírito santo	01
Cities	
São paulo	09
Salvador	07
Belo horizonte	06
Brasília	05
Recife	05
Rio de janeiro	04
Goiânia	03
Fortaleza	02
Porto alegre	02
Vitória	01
Bélem	01

(continued)

(continued)

Theses of Brazilians in France according to approach/location	
Curitiba	01
Teresina	01
Macéiio	01
Campinas	01
Olinda	01

Themes studied	
Urban geography, urban transport	
City and region, industrial/urbanization	
Urban policy, social movements, urbanism	
Tropical, popular housing, industry	61
Agrarian geography, expansion of agricultural frontier	30
Irrigated perimeter, dams	
Regional geography, spatial organization, *Amengement*	27
Geography of population/demography	11
Technical/scientific space	05
Region/ecology	04
Transport	03
Others	11

The tables mentioned above contain numbers that express the Brazilian spatial dynamic through themes that adjust themselves to the conceptual dominance of the time when the thesis was written. Brazil taken as a whole was studied in 31 theses with more general themes. In regional treatments, the northeast is dominant and then the Amazon. In the approach by state, Bahia is in first place, as the object of 15 theses. Next is Minas Gerais with 5 theses and then Paraíba with 4. As to the urban approach, São Paulo ranks first as the theme of 9 theses, followed by Salvador with 7 and Belo Horizonte with 6. The urban approach covers most of the Brazilian cities.

The general behavior of the postgraduate situation in geography in France shows a growing picture with an increase in the number of students and theses in the process of being elaborated. The thematic updating and the insertion of new thesis advisors confirm the dynamic of this process. The creation of study grants in the "sandwich" mode has intensified the quest for France in the process of forming Brazilian geographical professionals, but this mode causes difficulties in terms of statistical and research data and did not allow us to track their importance and position in the table that shows the general presence and production of Brazilians in France. We believe that there will be more interest and demand, amplifying the possibilities of exchange between two countries.

References

Berque J (1982) Recherche et cooperation avec Ie Tiers Monde, Rapport au Minisb'edela Recherche et de I'Industrie. La Documentation Française, Paris p 22

Velho O (1992) Impedindo ou criticando o processo de modernização? 0 caso do Brasil. In: Sfntese Nova Fase, vol 19, no 5, p 199

Chapter 8
Conclusions

Abstract Brazilian geography has an important role in the explanation of the reality of the country, Latin America, and why not the world. In Brazil, geographical science reached a reasonable level of reflection. The country has a considerable number of professionals who have achieved national and some international notoriety. In the context of the Mercosul, there is a wide space for discussion, interpretation, and analyses concerning the reality of the participating countries. The EGAL, Meeting of Latin American Geographers, has been a source of union among geographers. The events of the ANPEGE, National Association of Post Graduate Study and Research in Geography, and the AGB, Association of Brazilian Geographers, have attracted the attention and participation of geographers from neighboring countries. It is a good path for an effective integration.

Keywords Discussion · Interpretation · Analysis · Maturity

Geography as dynamic knowledge remains active and engaged, offering its contribution to a better understanding of a world that simultaneously ages and rejuvenates itself. Just as the world ages, geography also ages and rejuvenates itself developing new formulations, methods, and techniques capable of contributing to a better explanation of this increasingly complex world. Faced with the intricate situation, the world finds itself in the discussion about this or that school seems to lose meaning. In fact, the world seems more and more divided between the producers of science, society with its technical and scientific dominance and the dependents, and the consumers.

New spaces and new configurations of the world define themselves. The concept of a complex world gives the idea of unity, of coherence that while most often systemic does not seem able to give an account of reality in the face of increasing diversity on the face of the earth.

Ethnicities, languages, religions, and cultures guarantee the complex diversity in the way power and education are organized, the openness and closure toward the outside world. This world or worlds, unified and divided at the same time, have gone through significant changes in this century.

© The Author(s) 2016
J.B. da Silva, *French-Brazilian Geography*, SpringerBriefs
in Latin American Studies, DOI 10.1007/978-3-319-31023-7_8

The last years began with novelties: The fall of the Berlin Wall, the dismantling of the USSR, the coup in Russia, the serious economic crisis in Cuba, the rise of drug trafficking, the emergence of new power blocs, the dominion of social networks, smartphones, iPad, blue ray, and other technologies emerged in the market. The significant presence of China in the global scene, the retraction of the American empire. Globalization and the apparent unification of the market were without a doubt the most spectacular. Merchandize crosses the world, everything is exchanged, and everything is for sale. Flows and more flows of capital, persons, and goods are drawn on the world, crossing the globe in all directions. The advances in communication techniques reduce distances, without recognizing the artifice of territory, and convert the world into a large village. Immense structures are built to give a place for large events. They are ephemeral. They appear and disappear in the scenery, confounding observers. The transmission of competitions and events such as Formula 1 reach a formidable public. There are numerous identical programs, the Oscars, the Olympics, and the World Cup, among others. During the Gulf War, the video images of the CNN television network took on the characteristics of virtual images, appearing to corrupt notions and concepts of spatiality. Viewers anywhere in the world became involved and participated in the war. They followed the trajectory of the missiles; they entered and left the scene as though they were actors or players in the Woody Allen movie *Midnight in Paris.* Words such as warhead, fighter plane, Mirage jet, Exocet became part of everyday language. The world fits into enormous LCD screens. Events follow one another in life, death, party, state coups, stock exchange collapse, and massacres. Everyone took part in that war; everyone entered the field of operations. The real time of the fact is the time of the diffusion of the image.

In the midst of a modernity described by Habermas as "an unfinished project" came the struggle of minorities: among blacks, indigenous peoples, feminists, homosexuals, etc. The 1960s registered a change in values. The democratization of the body, sexual liberation and the environmental fight. The new map of the world with unprecedented configurations reminds us of the end of the Cold War which divided the world between the USA and the extinct USSR. This new configuration of the world on the economic level alters the so-called Third World which loses sense as a bloc. The Third World still gains visibility not for its importance but for the gigantic nature of its problems. Liberal winds reduce repression in Latin America alternated with short periods of democratization. The unraveling of colonial empires in Africa was followed by fragmentation into ethnic groups. In Asia, the emergence of China and India changes the profile of the continent altering flows, provoking concentrations and dense populations. Japan adapts with difficulty to this new Asian configuration.

Neo-Malthusianism revisited in several research centers, policy-making centers, and universities. Few countries joined the closed club of the rich. The demographic explosion happens on a large scale in the poorest areas of the world.

The partitioning resulting from the war machine and economic control is constant. Natural resources enter a period of exhaustion and impose care and a sharpening of environmental awareness. Energy dependency, based on petroleum, encourages research centers to seek new sources to keep the world in motion. The oil crises created a contradiction in some countries, which had strong revenues due to petroleum exports, but did not resolve their major economic and social issues.

The oil crisis generated a state of alert; it left dependent nations in an uncomfortable situation. The technology created and perfected in the two great wars of the century evolved on an unthought-of scale. The arms industry impelled the main blocs into a space race. The technological achievements were formidable. The new materials used and tested in space programs were put at the disposal of the market. Science advanced and with it the certainty of other possibilities. The branches of learning intersected in all directions; there was a complete immersion in the world of new technologies. Plastic artifacts, their counterparts, and other petroleum derivatives invaded all homes regardless of social class—a true revolution in everyday life. Their proliferation transformed the surface of the earth; their use replacing products considered classics caused controversy and rejection. This exacerbated environmental problems that took on the form of *issues*. As they are not biodegradable, petrochemical products create a treadmill of problems. In response, organizations and groups were established and mobilized, forming the *green movement* that proliferated rapidly and acquired various features, depending on the various forms of aggression toward nature and society. Today, the ecological issue follows different directions.[1] Scientists, politicians, and religious leaders stand against the excesses that lead to the depletion and extinction of natural resources.[2]

Earth, our well-known planet, insists on revealing new features and science continued acquiring new characteristics and trying new possibilities. Fine chemistry, computer science, and genetic engineering have introduced a new conception of science. Laboratories strive to solve new challenges, trying to reconcile technical advancement and the ecological issue. In this way, capital and science renew their alliances. Nevertheless, historical diseases such as syphilis and cholera persist. Others, such as cancer and AIDS, come into play. Hunger plagues the world acquiring epidemic features. Africa, the continent most in need of help, is the most abandoned.

The 1980s revealed the great crisis of capital. A new world order took shape so that the old model of identification of urban-industrial landscapes with a certain type of organization seemed unable to grasp the dynamics of this new type of spatial arrangement.

Powerful modern manufacturing has been transformed, it no longer has chimneys, and it does not occupy large blocks in unhealthy areas of big cities dominated by pollution, with a twilight ambience dominated by tones of gray. This new industry based on chips and fine chemicals can set itself up in any neighborhood of the metropolis, in technopoles. They no longer have assembly lines of workers with

---

[1]"In Latin America, for example, the degradation of popular national regimes in many cultures caused more than anything the triumph of military dictatorships and the exchange of protectionism for a liberal policy seeking comparative advantages in the global market..." IN: Touraine (1992).

[2]Ver Guattari (1990), Serres (1990), Ferry (1992), Morin (1973).

their shabby overalls; they no longer have smoky, unpleasant environments. Modern industry is confused with other urban forms, and it camouflages reality. Today, trade secrets of formulas and contracts animate executives, politicians, and scientists. The old order with its ancient, heavy space, full of pollution, is still needed and persists. Third World countries clamor for it and acquiesce to the technological disparity. The technopoles and strategic laboratories are indistinguishable in the urban landscape. Instead, they go unnoticed in a mimicry that hides their financial and industrial power.

> This is how the Silicon Valley[3] in the United States is a spontaneous construction, linked to cutting-edge technology and innovation created in laboratories, propelling the emergence of dynamic companies, attracting capital and expertise and producing an urban environment with its own features. The city that researches, finds and produces is a technopole, the engine and main link in a production system that acquires shape from it. It is celebrated as the birthplace of an agglomeration of companies with strong technological standards. The world's largest concentration of *brains* has a knock-on effect in terms of scientific processes. First of all, the intellectual environment plays a role in scientific creativity, information circulates better and new ideas are confronted with each other. Much more than this, innovations are often interconnected: setting up a particular business has induced innovation opportunities for others.[4]

Steeped in this upheaval that is shaking and transforming the world, geography, this respectable, centennial old lady, our old acquaintance, in her wisdom rejuvenates herself and is immersed in the search to understand reality in this time of rapid, sudden changes.

In his last book George (1990) *Le métier de geographe*, affirms:

> The dream of geographers at the start of the century was to draw an album of images in various scales to inform and edify future generations. By describing the world they felt they had fixed it.

He goes on to ask:

> The question is whether in this fast-paced adventure geography has not lost its identity.

He ends his book by saying:

> The circle is not closed as history is a variable spiral in a geometry of curves in which geography always has its place.[5]

Geography has built knowledge critical to the understanding of the Earth and the World. Each day this knowledge becomes more useful for a better comprehension of this world. On the threshold of the year 2000, geography, as a science, finds the breath to refresh and update itself and push past boundaries, abandoning outdated analyses, frayed concepts, and inadequate methodologies. This old world that renews itself every day imposes new challenges. The dynamics of this new era are

---

[3]Manzagol (1992, p. 510).

[4]Geledan and Bremond (1988b, p. 343).

[5]George (1990, p. 241).

intense, causing the emergence of newer new worlds. Understanding them by imprisoning or freezing them is impossible. It is up to geographers as professionals who deal simultaneously with nature and society, as subjects of the production process of space, to seek the renewal and adjustment of an entire theoretical–methodological arsenal to explain this reality and understand its dynamics.

The world today is a body of knowledge that has evolved gradually. From the disconnected early world, a kind of microworlds, we arrive at the global village. The market and communications command the process.

In this evolution, the world conceived as a field of forces, passes through the Cold War, and lives through capitalist expansion with the growing universalization of economic flows. Thus, they ordered and created hierarchies in world space. Modernity has been and remains the rationale for most political actions and government programs. According to Touraine (1992), "The exhaustion of the idea of modernity is inevitable because it is defined not as a new order but as a movement, a creative destruction."[6]

Geography faces major challenges at this start of millennium. There is no point discussing or asking whether there is a Brazilian school and which foreigners most influenced us. Specifically, we have a geography which such as others, questions its nature, its duality. However, when trying to deny all this, as a field of knowledge, it does not have enough firmness to settle with the other schools linked to other countries that have the same ambiguities with regard to their dichotomy.

Geography's originality lies in the fact that it addresses both nature and society, which must first and foremost ensure its unity and not divide it. Therefore, geography, in an integrated world, connected, at the end of the millennium, which in turn has its contradictions, should invest what it can to continue its scientific project, which is to explain society from geographical space, a product of the relationships established by society. As for the foreigners, those pioneering geographers who persistently laid the groundwork for the development of scientific geography in the country, who managed to establish qualified staff through training policies, who gradually replaced them. This substitution did not at any time mean giving up the contribution of foreign professionals. The exchange, interchange, and change are fundamental to the advancement of geographical science.

Among these foreigners, there is no doubt that the French occupied and occupy a special role. Relations between the two countries should be strengthened, allowing a reciprocal exchange. Today, at the start of the millennium, the possibilities for exchange are increased, and in the process, we are able to somewhat modify Brazil's role. We were apprentices; we inherited methodological procedures, theoretical lines, bibliography, etc. Over time, combining the most effective learning from the French and adding what we learnt with the Germans, Americans, British, and others, Brazilian geography has reached a particular way of doing geography. It is up to us as producers of science, emerging from this amalgamation, to demonstrate and disseminate it.

---

[6]Touraine, A. op. cit.

The interviewees were categorical in recognizing the autonomy and quality of Brazilian geography. Sachs, continuing his interview, said:

> I think revitalization and a debate on Josué de Castro's method in *The geography of Hunger* is a problem of international and not only Brazilian concern. Ab'Saber, I think today he is extremely important in the debate about the Amazon, for his ability to combine conformity to geography in the strict sense of the word with sensitivity to emerging problems. I've read his work and talked to him and it confirmed he is one of the top rank but the problem is that he is little translated… Orlando Valverde is a personality; he is from another generation and inspired a whole generation. I think Valverde's importance was the role he played as a man, rather than his books.

Brazilian geography has an important role in explaining the reality of the country, Latin America, and why not the world. We have achieved a reasonable level of reflection; we have a considerable number of professionals who have achieved national and even international prominence. In the context of the MERCOSUR, there is much to speculate about, discuss, interpret, and analyze regarding the reality of the countries involved. EGAL—The Meeting of Latin American Geographers has been a mark of the union between geographers. The events held by ANPEGE and AGB have attracted the attention of our neighboring geographers which is a good way forward for effective integration.

The history and reality of everyday life have proven how much has been produced in our country's geography. In this final phase of this research, we will give voice to our interviewees, for them to express their views on Brazilian geography:

> "… Brazil is not alone, but it is certainly one of the best countries studied internally, precisely because of the quality and vitality of its geographical school." Pierre George in a written response to the questionnaire in November 1992.

> "I believe that Brazilian geography has its own profile and autonomy." Jean Labasse in a written response to the questionnaire in December 1992.

> "Clearly, Brazilian geography exists. It has the same epistemological problem that geography has in every country…" Bernard Kayser in a written response to the questionnaire in November 1992.

> I think, in fact, that Brazilian geography has its own profile: after having engaged a lot in quantitative models in the 1970s, it reintroduced the political and the cultural dimensions in the analyses. And it manipulates computerized cartography easily. Places like USP and UFRJ seem to me at the same time, creative and autonomous. (Helene De Rivière D'Arc in a written response to the questionnaire, December 1992.)

> "It seems to me that Human, Economic and Social geography, have acquired a certain autonomy from French geography and other dominant schools." Jacques Malezieux in a written response to the questionnaire, December 1992.

Finally returning to Milton Santos, the eminent Brazilian geographer of international standing, who in answering the question about his relationship with France, today gave us an affidavit that, we believe, expresses the feeling that permeates our exchange in recent years:

France continues to have great importance in my life, in my career, in my ideas. It is a
contradictory relationship, not always peaceful inside me but it is extremely important
because I feel good here (in France), I feel pretty much at home and it is clear that this
creates problems. It is the same contradictory relationship that you have with your own land
which is also contradictory, the relationship I have with Bahia that is also contradictory.

References

AFDG (1991) Géographes Associés. n. 9. Université de Lyon II, Lyon
AGB (1956) Anais da Associação dos Geógrafos Brasileiros. São Paulo, AGB, v. VIII, i Torno I,
    406 p
Allies P (1980) L'Invention du territoire. Presses Universitaires de Grenoble, Grenoble, 184 p
Amorim Filho O (1985) Reflexões sobre as tendências teórico-metodológicas da Geografia.
    Instituto de Geociências, Belo Horizonte
Association Géographie et Cultures (1992) Géographie et Cultures. Paris, Ed. L'Harmattan, 144 p
Associação dos Pesquisadores e Estudantes Brasileiros na França. Nouvelles APEB, Paris, APEB,
    n. 7, dec. 1992 e n. 8, jan. 1993
Aubree M et al (1991–92) Cahiers du Brésil Contemporain. Maison des Sciences de l'Homme,
    Paris
Bailly A, Ferras R, Pumain D (1992) Encyclopédie de Géographie. Economica, Paris, 1132 p
Berque A (1990) Médiance de milieux en paysages. GIP RECLUS, Montpellier, 163 p
Boudeville JR (1973) Os espaços econômicos. Difusão Européia do Livro, São Paulo, 123 p
Bruyne P et al (1977) Dinâmica da pesquisa em ciências sociais. Francisco Alves, Rio de Janeiro,
    252 p
Bulletin de la Société des Professeurs d'Histoire et de Géographie de l'Enseignement Public.
    Historiens et Géographes, Neuilly, n. 216, rev. 1969, 583 p
Cardoso LC, Martiniere G (1984) France-Brésil, vingt ans de coopération. IHEAU PUG, Paris
Christofoletti A (1982) Perspectivas da Geografia. DIFEL, São Paulo, 318 p
CNRS-INTERGEO (1992) L'Annuaire de la recherche géographique francophone. INTERGEO,
    Paris, 862 p
Comission de Geografia (1968) Instituto Pan-americano e História. Simpósio de Geografia Urbana.
    Inst. Pan-americano de Geografia e História, Rio de Janeiro, 324 p
Comité National Français de Géographie (1984) La recherche géographique française. Paris-
    Alpes, CNRS- XXV Congres International de Geographie, 261 p
CRNS/INTERGEO (1966) Bulletin(s) INTERGEO, Paris, n. 1 a 104, 1966 a 1981
de Azevedo A (1970) Brasil a terra e o homem. Cia Editora Nacional, São Paulo, 490 p
Demangeon A (1974) O continente brasileiro. Difusão Européia do Livro, São Paulo, 191 p
Dollfus O (1972) O espaço geográfico. Difusão Européia do Livro, São Paulo, 121 p
Dollfus O (1973) A análise geográfica. Difusão Européia do Livro, São Paulo, 130 p
Droulers M (Coord) (1990) Le Brésil à l'aube du troisième millénaire. IHEAL/CREDAL, Paris,
    151 p
Dulles JWF (1985) O comunismo no Brasil. Nova Fronteira, Rio de Janeiro, 366 p
Durand M-F et al (1992) Le Monde, Espace et Système. Presses de la Fondation Nationale des
    Sciences Politiques/DALOZ, Paris, 565 p
Febvre L (1970) La terre et l'evolution humaine. Ed. Albin Michel, Paris, p 444
Fernandes AM (1990) A construção da ciência no Brasil e a SBPC. Ed. UnB, Brasília, 292 p
Ferry L (1992) Le nouvel ordre écologique. Grasset, Paris, 275 p
Foucault M (1986) A arqueologia do saber. Forense Universitária, Rio de Janeiro, 239 p
Fremont A (1988) France-Géographie d'une societé. Flamarion, Paris, 290 p
Galeano E (1986) As veias abertas da América Latina. Paz e Terra, Rio de Janeiro, 307 p

Gallais J et al (1991) Sahel, Nordeste, Amazonie. Ed. L'Harmattan/UNESCO, Paris, 233 p

Gebran P (Coord) Conceito de modo de produção. Paz e Terra, Rio de Janeiro, 275 p

Geledan A, Bremond J (1988a) Dicionário das teorias e mecanismos econômicos. Livros Horizontes, LDA, Lisboa, 463 p

Geledan A, Bremond J (1988b) Dicionario das Teorias e Mecanismos Economicos. Livros Horizontes, Lisboa, p 343

George P (1990) Le métier de géographe. Armand Colin, Paris, p 241

George P et al (1966) Geografia ativa. Tradução de Gil Toledo Manuel Seabra, Nelson de la Corte e Vicenzo Bochicchio. DIFEL, São Paulo, 359 p

George P et al (1979) Populações ativas. DIFEL, Rio de Janeiro, 219 p

Gourou P (1948) Les pays tropicaux. Presses Universitaires de France, Paris, 271 p

Gramsci A (1987) A concepção dialética da história. Civilização Brasileira, Rio de Janeiro, 341 p

Guattari F (1990) As três ecologias. Papirus, Campinas, 85 p

Harvey D (1992) Condição pós-moderna. Loyola, São Paulo, 349 p

IBGE (1963) Visita de mestres franceses. IBGE, Rio de Janeiro

IHEAL (1991) Cahiers des Amériques Latines. IHEAL, Paris, 143 p. Instituts et Centres de Recherches de Géographie. INTERGEO—Bulletin. Paris Laboratoire de Communication et de Documentation en Géographie (CNRS)- n. 1 a 104, anos de 1966 a 1991

Japiassu H (1978) Nascimento e morte das ciências humanas. Francisco Alves, Rio de Janeiro, 262 p

Labasse J (1973) La organización del espacio. Ins. de Estudios de Administratacion Local, Madrid, 752 p

Lacoste Y (1975) Geografia do subdesenvolvimento. DIFEL, São Paulo, 265 p

Lacoste Y (1981) Os países subdesenvolvidos. DIFEL, São Paulo, 120 p

Lacoste Y (1984) Unité et diversité du tiers monde. Ed. La Decouverte, Hérodote, Paris, 563 p

Lacoste Y et al (1979) Hérodote. Maspero, Paris, n. 16, oct/dec 1979

LARES (1991) Les raisons de l'Urbain- Colloque international. Rennes. LARES, Université Rennes, 2, 347 p

Ledrut R (1983) Espaces et societés. Ed. Anthropos, Paris, n. 42, jan/juin 1983

Lefort C (1990) As formas da história, 2. ed. Brasiliense, São Paulo, 345 p

Levy J (1991) Géographies du politique. Presses de la Fondation Nationale des Sciences Politiques et Espaces Temps, Paris, 220 p

Manzagol C (1992) La Localisation de activitesspecifiques. In: Encyclopedie de Geographie. Economica, Paris, p 510

Marrero L (s.d.) Viajemos por el mundo. Publicaciones Cultural, S.A., Cuba, 240 p

Merquior JG (1987) O marxismo ocidental. Nova Fronteira, Rio de Janeiro, 323 p

Monbeig P (1975) O Brasil. DIFEL, São Paulo, 133 p

Monteiro CAF (1981) A questão ambiental no Brasil—1960–1980. USP/Inst. de Geografia, São Paulo, 133 p

Moraes ACR (1981) Geografia: pequena história crítica. HUCITEC, São Paulo

Morin E (1973) Le paradigme Perdu: La Nature Humaine. Seuil, Paris

ORSTOM (1986) Cahiers des Sciences Humaines. ORSTOM, Paris, v. 22, n. 3/4, 480 p

Raffestin C (1980) Pour one géographie du pouvoir. Librairies Téchniques, Paris, 249 p

Reclus E (1985) Geografia. Ed. Atica, São Paulo, 200 p

Reynaud A (1986) O espaço interdisciplinar. NOBEL, São Paulo, 139 p

Ribeiro ACT et al (Org) (1990) Metropolização e rede urbana-perspectivas dos anos 90. IPPUR/UFRJ, Rio de Janeiro, 263 p

Rivière D'Arc H et al (1990) A Amazônia na França. St. Just-La-Pendue, CHI'T, 63 p

Roncayolo M (1990) La vine et ses territoires. Gallimard, Paris, 273 p

Santos M (1965) A cidade nos países subdesenvolvidos. Ed. Civilização Brasileira, Rio de Janeiro, 175 p

Serres M (1990) Le Contrat Naturel. Editions Francois Bourin, Paris

Touraine A (1992) Critique de la Modernité. Fayard, Paris

Printed in the United States
By Bookmasters